REDUCING THE GREENHOUSE GAS EMISSIONS OF WATER AND SANITATION SERVICES

*Overview of emissions
and their potential reduction
illustrated by utility know-how*

REDUCING THE GREENHOUSE GAS EMISSIONS OF WATER AND SANITATION SERVICES

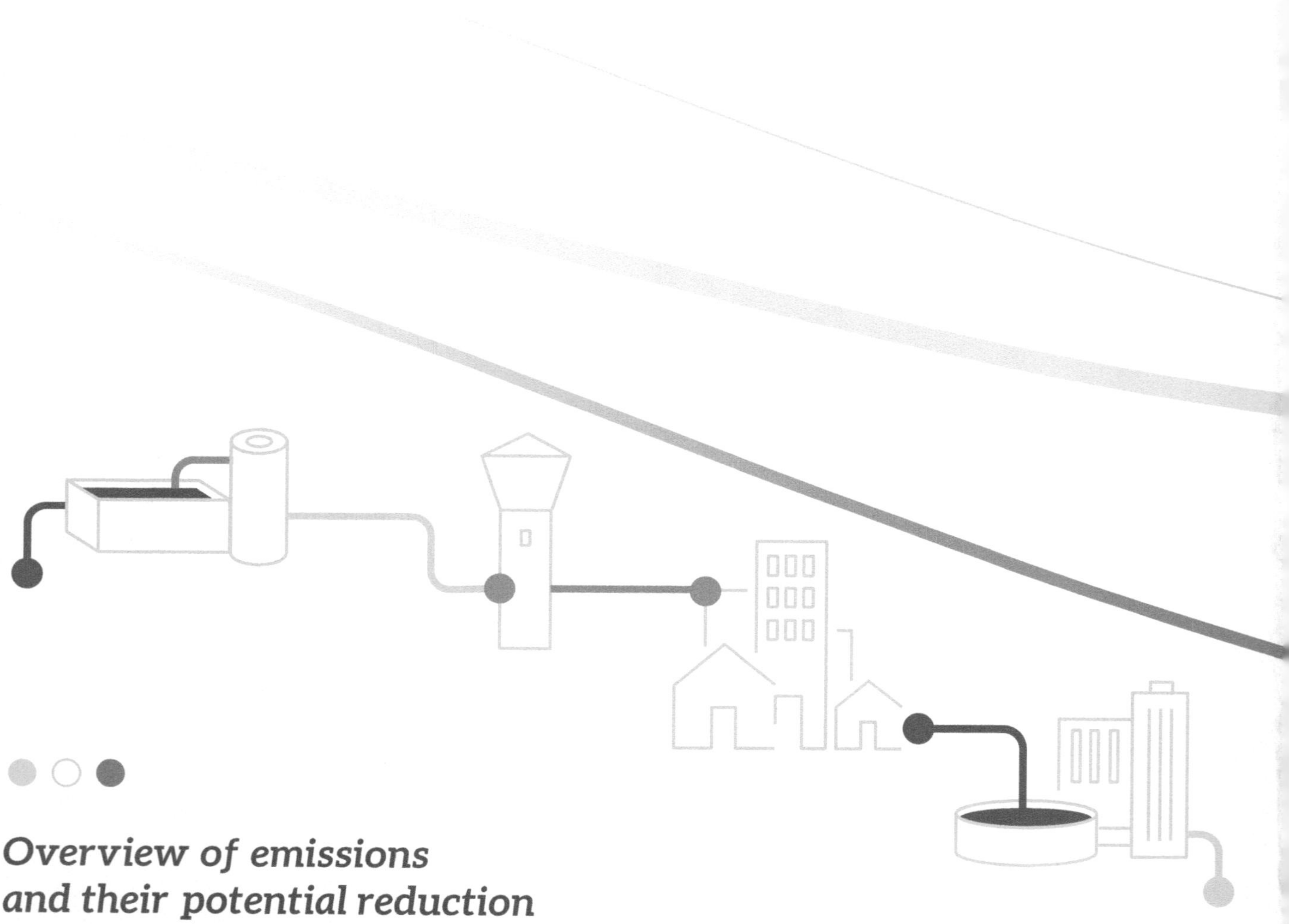

*Overview of emissions
and their potential reduction
illustrated by utility know-how*

Co-published by IWA Publishing,
Unit 104-105, Export Building, Republic,
1 Clove Crescent, East India,
London E14 2BA, UK

Tel. +44 (0) 20 7654 5500
Fax +44 (0) 20 7654 5555
publications@iwap.co.uk
www.iwapublishing.com

ISBN: 9781789063172 (eBook)
DOI: 10.2166/9781789063172

EDITORIAL NOTE

Why this report?

An increasing number of cities both in the Southern and the Northern hemisphere are experiencing chronic water stress. For instance, California has been experiencing a year of unprecedented drought, and Chennai in India reached "day Zero" in July 2019, meaning that, by that day, the city had used up all the water available through local resources. Climate change will make these situations increasingly frequent across the world, as evidenced in the map of water stress drawn up by the World Resources Institute (WRI)[1]. Water is one of the main indicators of climate change and changes in water regimes will require many adaptation measures for all sectors that depend on water resources. The French Water Partnership (FWP) has gathered in this publication an inventory of such solutions implemented by French operators. Members of the International Water Association's Climate Smart Utilities (CSU) Community of Practice have contributed to this report as reviewers, with the aim to support urban water utilities on their journey to adapt, mitigate, and be climate leaders.

The treatment and transport of drinking water and wastewater represents between 1 and 18% of electricity consumption in urban areas.

Achieving sufficient amounts of good-quality water available for users relies not only on good management of water resources, but also on technologies, such as desalination, which require significant additional energy. To reach the Sustainable Development Goals (SDGs), we must explore ways to reduce our pressure on natural resources, in particular by improving the efficiency of production processes. All this despite the increasing needs that come with population growth.

The treatment and transport of drinking water and wastewater represents between 1 and 18% of electricity consumption in urban areas[2]. As presented in this report, the water sector can contribute to reducing greenhouse gas (GHG) emissions from energy, but also from direct and indirect emissions associated with the activities that deliver the service.

Following the overview of where emissions come from, the report presents a review of proven measures to reduce GHG emissions which can be structured around the following three pillars:

1. Implementing **lean and efficient** approaches: reducing leaks, saving water and energy, etc.

2. Embracing **circular economy**: recycling wastewater, producing energy on-site, etc.

3. **Decisions and Strategy** to support the transition to carbon neutrality: purchasing green energy and low carbon supplies, training, awareness, or adapted governance structures.

These solutions are illustrated with concrete examples implemented by or for local authorities.

It is important to note that, in addition to solutions implemented by water services, consumers can play their part in reducing emissions associated to the urban water cycle through reducing consumption, water heating. For instance, significant progress in housing and the design of appliances have resulted in reduced household water needs. Consumers can also have an impact on wastewater treatment energy requirements through a better understanding of what belongs in the wastewater pipes, in particular in regards to micropollutants.

This report offers an overview of GHG emissions and of solutions to reduce them, and illustrates these with examples of French or international know-how. In France we observe that drinking water needs have stabilized over the past three decades although population has increased and household services have improved. This goes to show how useful it is to optimize all the aspects that can improve service efficiency, including GHG emissions. As services now focus on adapting to the impacts of climate change, the opportunity to also reduce the emissions needs to be on top of our minds to ensure we play our part and inspire others to play their part on the journey to carbon neutrality.

The solutions illustrated in this report are each specific to their context, and are intended as an inspiration for the readers to rethink solutions in their specific context. It is a toolkit which we hope will be useful both in urban and rural environments for local officials and those in charge of designing and managing these services which are essential for human needs.

The solutions illustrated in this report are each specific to their context, and are intended as an inspiration for the readers to rethink solutions in their specific context.

CO-SIGNED BY:

Jean Launay
President of the French Water
Partnership

Kala Vairavamoorthy
Executive Director of the
International Water Association

Jean-Luc Redaud
President of the FWP working group
on "Climate Change and Global
Environmental Changes"

1. WRI Water Risk Atlas Beta Aqueduct
2. Marc Florette, Léon Duvivier, Eau et Énergie sont indissociables,

The French Water Partnership (FWP) is the reference platform for French public and private water stakeholders who wish to be active at the international level. For the past 15 years, it has been advocating internationally to make water a priority in sustainable development policies, promoting French know-how, practices and experiences, and facilitating exchanges between French and foreign actors. The FWP promotes the collective messages on water established by its members (public institutions, local authorities, NGOs, private sector actors, but also research institutes and individual experts) at international forums such as World Water Forums, Climate and Biodiversity COPs, and high-level political forums on the Sustainable Development Goals.

The International Water Association (IWA) is the network of water professionals striving for a world in which water is wisely, sustainably and equitably managed. Drawing exceptional professionals from more than 140 countries, the membership of IWA brings together scientists, researchers, technology companies, water and wastewater utilities, and wider stakeholders involved in water management. Focused on the global water sector, the Association was launched in its current identity in 2000, building on a 70-year heritage of connecting water professionals around the world.

IWA recently launched the **Climate Smart Utilities** Initiative to assist water, wastewater, and urban drainage companies in improving their resilience by adapting to a changing climate while contributing to significant and sustainable reduction of greenhouse gas emissions.

REDACTION AND COORDINATION: Alexandre ALIX (French Water Partnership), Laurent BELLET (EDF), Corinne TROMMSDORFF (Water Cities), Iris AUDUREAU (French Water Partnership)

STEERING COMMITEE: Alberto AQUISTAPACE (Solidarités international), Camille ARNAULT (Agence de l'Eau Rhône Méditerranée Corse), Guillaume AUBOURG (pS-Eau), Vincent CASTAGNET (Up to green), Jean COMBY (ESF), Muriel DESGEORGES (ADEME), Sébastien DUPRAZ (BRGM), Caroline ORJEBIN (Suez), Jean-Eude MONCOMBE (WEC), Stéphane POUFFARY (Energie 2050), Jean-Luc REDAUD (French Water Partnership), Catherine THOUIN (CHF)

GRAPHIC DESIGN BY: Anne-Charlotte DE LAVERGNE (ancharlotte.fr)

WE ALSO THANK THE FOLLOWING PEOPLE FOR THEIR CONTRIBUTIONS TO THIS PUBLICATION: Léo BOUSQUET (Eau de Paris), Caroline CHAL (SYCTOM), Catherine CHEVAUCHE (Suez), Adeline CLIFFORD (ASTEE), Mathilde DENIAU (Veolia), Maxime FANTINO (ALEC Montpellier Métropole), Elodie FROSSARD (Association Migrations et développement), Jérémie JAEGER (Ville de Paris), Laura LEFLOCH (SIF), Solène LEFUR (ASTEE), Tristan MILOT (SIAAP), Camélia MORARU (French Water Partnership), Armelle PERRIN-GUINOT (Veolia), Patrick SAMBARINO (ESF), Marie-Laure VERCAMBRE (French Water Partnership)

FINAL REVIEW: Chapter 1: Jose PORRO (Cobalt Water), Martin KERRES (GIZ) / Chapter 2: Steven KENWAY (University of Queensland), Gustaf OLSSON (Professor Emeritus) / Chapter 3: Amanda LAKE (Jacobs), Martin SRB (Prague Wastewater Utility PKV), Marie RØDSTEN SAGEN (Bergen Kommune) / Chapter 4: Martin KERRES (GIZ), Katharine CROSS (Consultant)

MAY 2022

Table of contents

Towards a climate-change mitigation strategy for water and sanitation services

MAY 2022

The scientific evidence contained in the three parts of the 6th Report (AR6) of the Intergovernmental Panel on Climate Change (IPCC) is another reminder of the urgent need to respect the 2015[3] **Paris Agreement.** 195 countries agreed to the goal of limiting long-term global temperature increase to "well below 2°C" compared to pre-industrial levels and to pursue efforts to limit the increase to 1.5°C by massively reducing their emissions of carbon dioxide and other greenhouse gases (GHGs).

Climate change particularly affects water and it is still difficult to measure what impact exceeding the 2°C limit would have. All sectors of society are affected by the need to decrease greenhouse gas emissions. Industry ad hock estimates suggest that **emissions by water and sanitation services represent between 3 and 7% of cities' GHG emissions[4]**. As a comparison, the aviation sector causes 2 to 3% of global GHG emissions. Emissions from the water sector result as much from the non-treatment of wastewater and the pollution of aquifers in less developed countries as from those due to the operation of treatment plants. To address this, the **IPCC** suggests mitigation possibilities such as **converting wastewater into energy, reducing the consumption of water and energy** as well as other actions that fit into a circular economy approach.

One characteristic of water and sanitation services is that they are interconnected with many other sectors such as energy, agriculture, goods and services production and waste. As a result, the greenhouse gas emissions of water and sanitation services may affect, or be affected by, these other sectors. For example, after heavy rainfalls, waste accumulating in streets may clog collection networks, causing floods, and these results in additional energy use to clean the affected rainwater collection networks and to treat the waste and pollution thus generated. This indirectly increases the greenhouse gas emissions of sanitation services.

France's National Low Carbon Strategy (SNBC) has defined a roadmap to reach carbon neutrality by 2050 (also an EU target). This is in line with the European Union's Green Deal goal of reducing net greenhouse gas emissions "by at least 55%" below 1990 levels by 2030. Mitigation strategies are essential to be able to meet these ambitious targets.

Water and climate questions are usually addressed from the perspective of adaptation to climate change. For the water sector the mitigation aspect has been less studied up till now.

This report fits into the broader context of the interdependence of energy and water (Water-Energy Nexus), rather than the better-known water requirements for the energy sector, recognising that the need to reduce the GHG emissions of water and sanitation services goes with the growing demand for water. It is anticipated to increase[4] by 50%between now and 2030 worldwide, and by 70% for drinking water, due to the combined effects of population growth, economic development and the shift in consumer patterns. The drinking water and sanitation services treated in this study represent a non-exhaustive part of the water services in the city (e.g. stormwater management).

This synthetic report aims to provide an overview of possible levers to reduce the greenhouse gas emissions of water and sanitation services and provides an analysis of how adaptation measures can embrace this low-carbon approach.

---

3. Le Monde 23 juin 2021 Dérèglement climatique : l'humanité à l'aube de retombées cataclysmiques, alerte un projet du rapport du GIEC
4. Global Water Crisis – The Facts, UNU/INWEH, 2017

Reducing GHGs from water and sanitation services

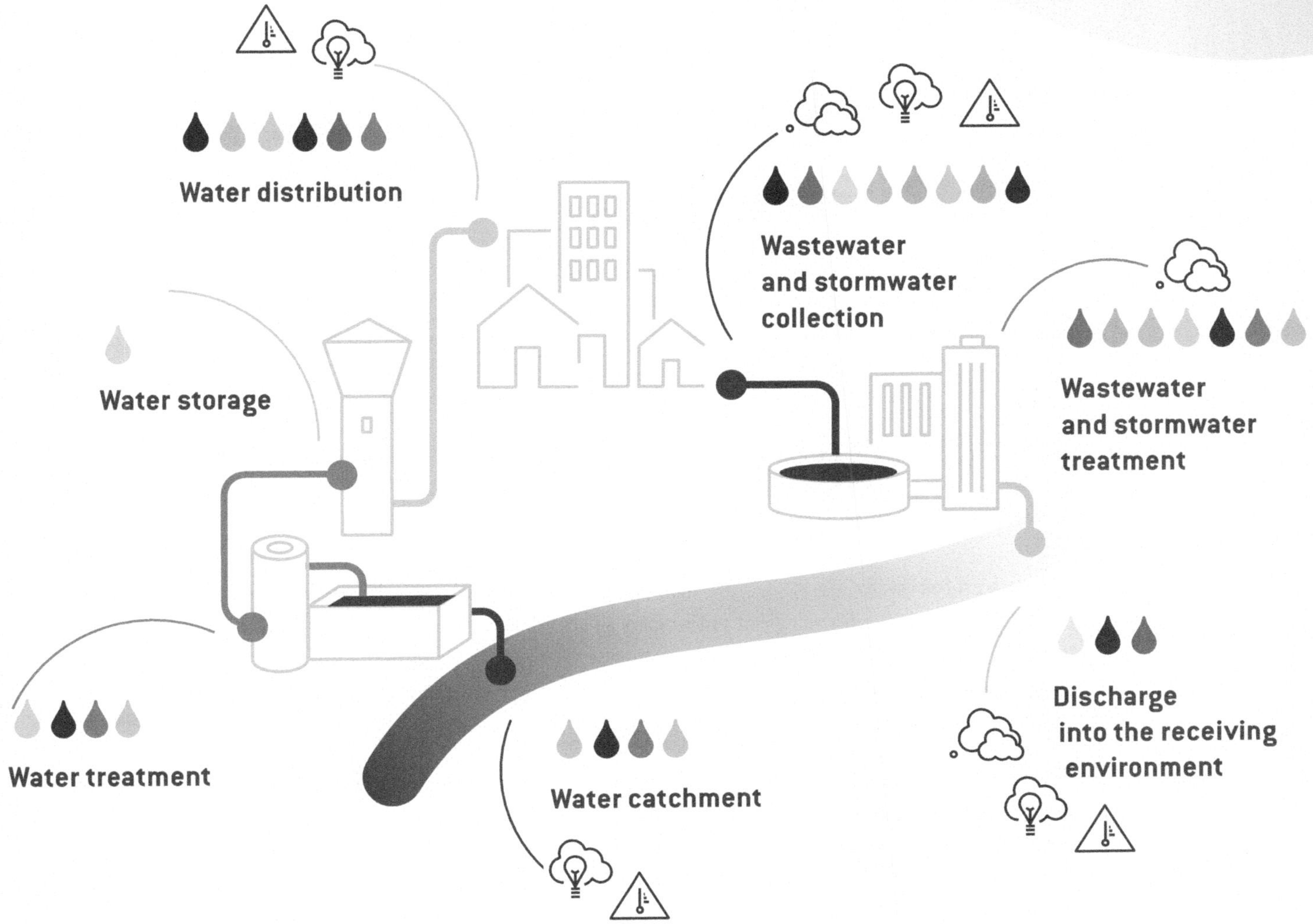

LEGEND

RISKS

- CH_4, N_2O, Co_2 not energy-related
- Co_2 energy-related
- Climate-related risks

LIST OF MITIGATION ACTIONS

SOBRIETY

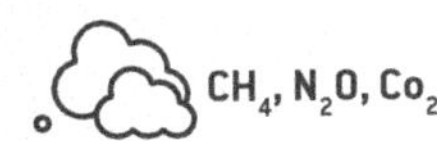

- Reducing water loss and infiltration
- Efficiency of services
- Reducing water and water-related energy consumption by end users
- Low-impact wastewater and rainwater treatment
- Alternative rainwater management
- Restoring and preserving the quality and quantity of water resources

CIRCULAR ECONOMY

- Reusing water, nutrients and materials
- Using available land and surfaces to produce solar and wind energy
- Producing heat
- Producing hydro-electricity
- Producing energy from biosolids

STRATEGIC CHOICES

- Awareness raising and education
- Governance that supports changing practices
- Economic incentive for responsible consumption
- Choosing low-carbon energy and supplies

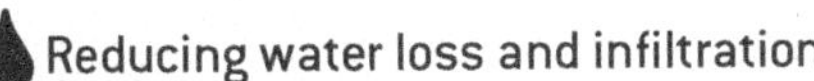

GHG emissions *of* water and sanitation services in a context of global changes

1. WHAT IS A CARBON FOOTPRINT VERSUS GHG EMISSIONS?

The **carbon footprint** measures the amount of greenhouse gases (GHGs) related to any human activity. It depends on the direct emissions associated to the activity, on the energy used to support this activity, in particular when fossil resources are used, as well as on the supply chain footprint which corresponds to the inputs and outputs of the activity, along with their transport and their respective GHG footprint during their production and use.

GHG emissions are categorized into 3 "scopes" by the international **GHG protocol: direct emissions** resulting directly from the activity (Scope 1), **indirect emissions** associated with the energy required by the activity (Scope 2) and indirect emissions related to the activity but caused by other organizations (Scope 3). This publication will focus mostly on Scope 1 and 2 emissions, and mention some Scope 3 emissions, acknowledging that the approaches to assess these latter emissions are not yet standardized.

The carbon footprint of an activity is a way of measuring its impact on the world's climate and to identify the activities that emit most GHGs.

In other terms, the carbon footprint of an activity is a way of measuring its **impact** on the world's climate and to identify the activities that emit most GHGs. The figure thus obtained serves as a tool to drive decisions concerning strategies and solutions to reduce the GHG emission of the said activity.

2. WHAT ARE THE SECTOR'S GHG EMISSIONS?

Emissions from water and wastewater services come from energy and direct emissions of nitrous oxide and methane during wastewater management.

Water and wastewater management are energy intensive processes, causing significant carbon emissions whenever fossil energy sources are used. The International Energy Agency estimated that in 2014, processes associated with abstracting, supplying and treating water and wastewater accounted for around 4% of the international electricity consumption[5]. When residential water heating is included, the total energy requirements rises to 3-5%[6] of total primary[7] energy use. Globally there is a lot of inconsistency in the reporting of what constitutes "water-related energy". Part of the confusion is due to ambiguous terms such as "system", "supply" and "sector". Recent analysis indicates up to 12.6% of total national primary energy use can be influenced by water, when both water-related energy of water users, and the energy use by water utilities, are included. Water heating for residential, commercial, and industrial purposes is the dominant fraction. Water and wastewater utilities use 0.4-2.3% of primary energy or 0.6-6.2% of regional electricity, mostly for water pumping.[7]

> *Figure 1:* **Water related energy as a percentage of total primary energy consumption across each country or region.**[7]

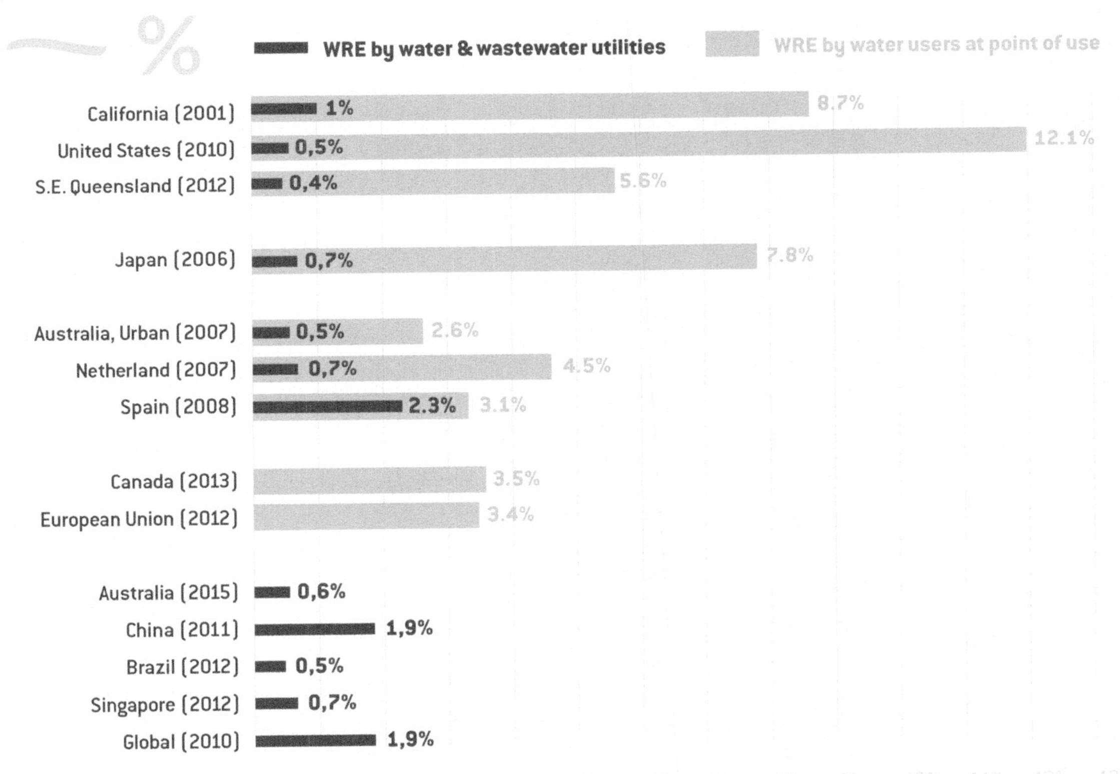

5. IEA (International Energy Agency) (2016): Water Energy Nexus. Excerpt from the World Energy Outlook 2016.

6. Kenway et. al. (2019). *Defining water-related energy for global comparison, clearer communication, and sharper policy.* Journal of Cleaner Production, 236 117502, 117502. doi: 10.1016/j.jclepro.2019.06.333

7. Total primary energy use covers consumption of the energy sector itself, losses during transformation (for example, from oil or gas into electricity) and distribution of energy, and the final consumption by end users of all types of energy (electricity, gas, petrol,…).

It is estimated that wastewater treatment in centralised facilities contributes alone some **3% of global nitrous oxyde emissions and 7% of anthropogenic methane emissions**

In addition, wastewater handling can be a source of the highly potent greenhouse gases methane and nitrous oxide. It is estimated that wastewater treatment in centralised facilities contributes alone some 3% of global nitrous oxyde emissions[8], and that methane emissions from wastewater collection and treatment processes contribute some 7%[9] of anthropogenic methane emissions. It was estimated in 2021 that 37% of urban wastewater worldwide was not collected, and 48% was not treated[10] , which results in a significant amount of nitrous oxide and methane emissions that could be reduced through proper management.

No global figure is scientifically consolidated regarding the contribution of urban water and wastewater services to global GHG emissions, however industry ad hoc estimates suggest this would be around 3% when including residential water heating. Again, based on industry adhoc estimates, potential for emissions from urban water and wastewater services can be ranked in order of importance as follows: 1/ end users water heating, 2/ direct emissions from nitrous oxide and methane associated to wastewater management, 3/ energy consumption for drinking water supply from abstraction to treatment (mostly pumping), and 4/ energy consumption for wastewater management.

Below is an overview of the type of GHG emissions from water and wastewater services, detailed under each of the three scopes of the GHG Protocol.

DIRECT EMISSIONS (SCOPE 1)

Emissions caused by wastewater and sewage sludge treatment

Emissions associated with the treatment process vary depending on the level and type of treatment. The GHGs emitted are caused by the degradation of organic matter and nitrogen present in wastewater. It produces two types of emissions:

- **methane (CH_4) emissions** which occur when water carrying heavy amounts of organic matter (carbon chains) remain in anaerobic conditions. Methane has a warming potential about 30 times higher than carbon dioxide over a 100-year timescale, but it is 80 times higher over 20 years[11] , which led to the COP26 agreement's proposal of focusing efforts on reducing methane emissions by 2030 to curb global warming.

- **nitrous oxide (N_2O) emissions** caused by the degradation of nitrogen compounds present in water under aerobic or anoxic conditions. Nitrous oxide's warming potential is estimated 300 times higher than carbon dioxide's over a 100-year timescale.

8. Law et. al. (2012) reported these figures for US and UK : https://www.ncbi.nlm.nih.gov/pmc/articles/PMC3306625/

9. GlobalMethane.org Factsheet (2013): Municipal Wastewater Methane: Reducing Emissions, Advancing Recovery and Use Opportunities (https://www.globalmethane.org/documents/ww_fs_eng.pdf)

10. Jones, E., *et al.* (2021). Country-level and gridded estimates of wastewater production, collection, treatment and reuse https://essd.copernicus.org/articles/13/237/2021/

11. McKinsey Report " Curbing Methane Emissions" September 2021, Pages 21-22

INDIRECT EMISSIONS (SCOPE 2)

Those are the emissions caused by the production of energy by a third party supplying energy to the water and wastewater services company. These emissions are usually assessed using the energy mix of the country and will therefore be higher if the share of fossil fuels in the country's energy mix is large. Offsets can be calculated when renewable energy is specifically purchased.

The following chapter provides an overview of the energy used by water and sanitation services worldwide. Figure 2 below gives a snapshot of the links between CO2 emissions and the urban water sector.

INDIRECT EMISSIONS (SCOPE 3)

These emissions are mainly caused by:

- the construction of capital assets or depreciation (construction and use of goods such as buildings, pipes, infrastructure, etc.),
- the necessary chemical products and reagents,
- the reuse of by-products such as sludge (composting, spreading, etc.),
- discharge into surface water.

Among indirect emissions, those specifically associated with infrastructure investments are too often overlooked and the water and sanitation services' investment needs are particularly high in developing countries. ADEME, the French Energy Transition Agency, suggests an evaluation method based on accounting for assets, which spreads the GHG impact of investments out over their lifespan.

> *Figure 2:* **Link between urban water services and CO_2 emissions**[12]

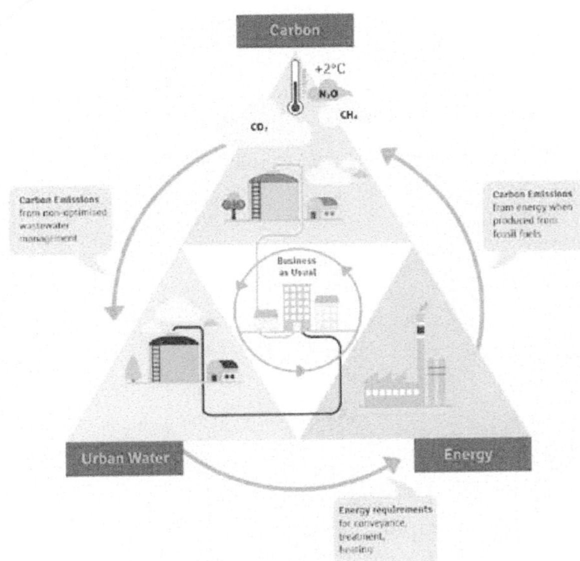

Business as usual aggravating the impacts of climate change

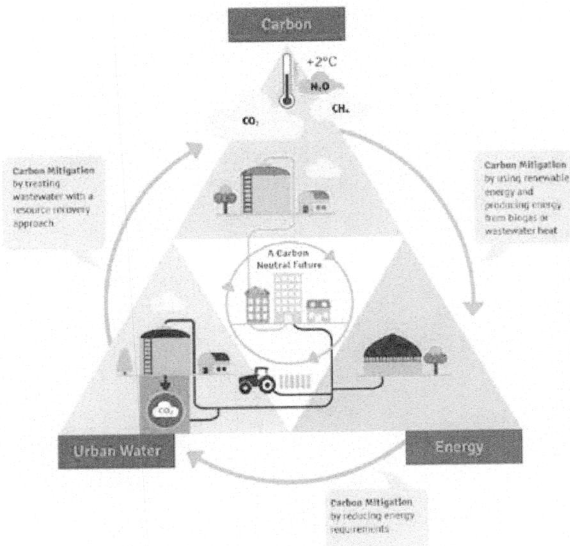

Low-carbon utilities for a carbon-neutral future

12. Source : The IWA Roadmap to a Low-Carbon Urban Water Utility
https://climatesmartwater.org/wp-content/uploads/2018/09/figure01_nexus_sidebyside-01.png

SPOTLIGHT : How technology choices impact GHG emissions of wastewater services

THE CHOICE OF SANITATION PROCESSES

The quantity of GHG emissions varies depending on the sanitation process in place. An INRAE[13] study of intensive processes reveals that technologies such as sequencing batch reactors (SBRs), membrane bioreactors (MBR) and moving bed biofilm reactors (MBBR) require more energy than conventional activated sludge (CAS) or biofilters. When choosing these technologies, planners should reflect on whether these more energy intensive technologies are really needed to meet requirements that traditional less energy-intensive processes could not meet. In the case of large plants requiring intensive processes, CAS tends to be more practical and less energy-intensive than say MBR. Beyond the energy requirement, some activated sludge processes may lead to higher nitrous oxide emissions. More on the nitrous oxide topic can be found in the IWA publication Quantification and Modelling of Fugitive Greenhouse Gas Emissions from Urban Water Systems (Ye et al., 2022).

For individual sanitation and small, decentralized plants (up to 2,000 equivalent residential unit), there are more **"rustic" treatments,** meaning extensive treatments like natural **lagooning** or **filtering gardens**. These processes tend to consume less energy than intensive biological or physical and chemical treatment processes, and when well-managed, will also produce low methane and nitrous oxide emissions

PROCESSING SLUDGE[14] :

Emissions associated with the processing and reuse of sludge also must be taken into account. Various processes can be used to reduce the water content and the volume of sludge, which is done in several steps. Solar-type drying, which achieves 60 to 80% desiccation, uses up between 30 and 100 kWh/dry ton. Other methods, whereby 20-30% desiccation may be achieved, are: liming treatments (10 kWh/TMS), dehydration with band filters (20 to 40 kWh/ dry ton) or plate filters (30 to 40 kWh/ dry ton), or centrifugation (60 to 80 kWh/ dry ton). Finally, among the processes that achieve 3 to 8% desiccation, i.e. that thicken sludge, energy requirements vary widely: low for gravity thickening (5 to 10 kWh/ dry ton), slightly higher for gravity drainage (25 to 60 kWh/ dry ton) and much higher for flotation thickening (100 - 130 kWh/TMS). The most energy-intensive thickening technique is centrifugation (150 to 200 kWh/ dry ton). Beyond requiring energy for thickening and dewatering, biosolids can be partially transformed into biogas to produce renewable energy used on site or sold. Sludge storage can result in unintended methane emissions and is a function of detention time.

When choosing sludge processing alternatives, planners should reflect on the impact they will have on GHG emissions in addition to the specific requirements of the plant process and subsequent sludge reuse options.

---

13. Lit fluidisé sur supports plastiques mobiles, Moving bed biofilm reactor

14. **IRSTEA et AERMC : Rapport final Consommation énergétique des filières intensives de traitement des eaux résiduaires urbaines, 2017** Study carried out on 310 installations in France

Water and sanitation services emissions vary depending on the context: the country's energy mix, the treatment process and operation, the quality and quantity of available water, the level of treatment required before discharge into the receiving waters, and whether renewable energy can be produced or purchased. For instance, the supply and opportunity to produce low-carbon energy will not be the same for towns located in mountains (pico- or micro-turbines pour hydroelectricity) or in plains (wind turbines).

Chapter 3 of this publication highlights enablers and solutions to reduce greenhouse gas emissions. It inspires us to take action at all stages of the water cycle and regardless of facility size.

3. WHY ASSESS GHG EMISSIONS?

As water and sanitation services are expected to massively invest in infrastructure over the next decade, both to meet SDG6 target on clean water and sanitation and to adapt to the impacts of climate change, it is critical that these services start measuring their GHG emissions (or consider low-carbon or less energy-intensive technologies for new infrastructure) and are able to show how their investments contribute to reducing and minimizing global GHG emissions. Beyond the contribution to national goals, the ability to measure, monitor and report is also an opportunity for Utilities to access green finance.

This chapter provides an overview of the context that drives need for GHG assessment and monitoring.

3.1 Water and sanitation services: what are the needs worldwide?

Water and sanitation services are heavily reliant on the availability of freshwater, which is unevenly distributed across the world. According to the FAO, annual freshwater resources are mainly located in South America and South Asia; they are weaker in Africa, the Middle East, Central Asia and Australia.

SDG 6 calls for universal access to sustainably managed water and sanitation services by 2030. At the current rate of progress, we are not going to reach this target. The number of persons without access to drinking water as defined in SDG indicator 6.1.1 was 2.2 billion in 2017[15] , which is only a 4% decrease since 2000. And in 2020, 2.1 billion people still had no access to drinking water as defined in SDG indicator 6.1.1, according to a joint report by the WHO and UNICEF[16] . The same report estimates that one person out of four lacked safely managed drinking water at home and nearly half the world population lacked access to safely managed sanitation services. If the progress rate is not multiplied by four, then in 2030, deadline for the Agenda 2030, billions of people still won't have access to safely managed drinking water, sanitation and hygiene services.

The growing demand for drinking water, demographical evolutions, as well as conflicts, climate change and urbanization are already posing problems for water supply systems. In 2020, on average, 10% of the global population lived in countries with high or critical water stress, which has a significant impact on water access and availability for personal needs[17] . The configuration of water and sanitation services varies widely from one part of the world to another. Centralized networks are most common in developed countries, while water services in developing countries are organized in more varied ways.

---

15. Progress on household drinking water, sanitation and hygiene 2000-2017. Special focus on inequalities. New York: United Nations Children's Fund (UNICEF) and World Health Organization, 2019

16. Progress on household drinking water, sanitation and hygiene 2000-2020: five years into the SDGs. Geneva: World Health Organization (WHO) and the United Nations Children's Fund (UNICEF), 2021

17. FAO and UN Water. 2021. Progress on Level of Water Stress. Global status and acceleration needs for SDG Indicator 6.4.2, 2021. Rome.

Indeed, in developing countries, the efficiency of conventional public networks (interconnected infrastructures centrally managed by a single operator) is challenged by instability and by growing demand for water on the part of low-income populations. That is why **alternative types of supply** have been developed either through public-private partnerships or offers outside networks (water sold from trucks, non-collective sanitation, etc.). Not all of these alternative approaches comply with the access definition used for SDG 6. In cities in emerging economies conventional systems in wealthy neighborhoods may exist alongside alternative systems, of lower efficiency and uncertain viability, in poorer neighborhoods and slums[18].

The reasons are partly to be found in the absence of land regulation in poor neighborhoods, which develop independently, by limited technical services, and by the lack of local taxation which creates a financing problem for water and sanitation services.

URBAN AREAS:

Urban areas and rural areas have unequal access to water. According to the WHO and UNICEF, nearly all **people using untreated surface water live in rural areas[19]**: 150 million vs. 11 million only in urban areas. Being wealthier and more dense, urban areas are equipped with more modern, better-quality sanitation systems than rural areas. In addition, rural populations have less access to latrines, and open defecation is a more common practice there. As urbanization progresses (UNICEF expects urban population will have grown by 50% in developing countries in 2050) essential services will need to be developed accordingly.

REGIONAL PERSPECTIVES

Many countries in **Africa** face a general lack of infrastructure to collect and treat wastewater. In many cases, this causes surface water and groundwater pollution, knowing water is often scarce in the first place. Existing systems cannot answer the other challenges facing the African continent, namely rapid population growth, rampant urbanization and increased demand for water.

In the **Middle East**, water availability is a major issue. The main solutions are decreasing water demand, e.g. by reducing water losses and complementing supply through desalination plants and the reuse of treated wastewater. In 2013, 71% of wastewater collected in Arab nations were treated again, and 21% of this volume used for irrigation or to replenish water tables.

In **Europe and North America**, 95% of the population has access to improved sanitation and treatment levels are constantly improving. But significant volumes of water remain untreated after collection, especially in Eastern Europe, although tertiary treatment has become more widespread.

In **Latin America and the Caribbean**, a mere 20 to 30% of wastewater is collected. Water and sanitation coverage is increasing. Some arid areas are still dependent on long-distance water transfers.

In parts of **Asia**, water and sanitation services are not very efficient, meaning efforts are required to use water more rationally. Competition for freshwater availability is already intense in Asia. In Pacific Asia, 80 to 90% of wastewater is still discharged without treatment, which results in significant pollution of groundwater and surface water. Relevant services will need to become resilient, as they are vulnerable to the rise in sea levels caused by climate change[20].

---

18. Naulet F., Gilquin C., Leyronas S., Eau potable et assainissement dans les villes du Sud la difficile intégration des quartiers défavorisées aux politiques urbaines GRET,2014

19. UNICEF OMS, Rapport Progrès en matière d'eau, d'assainissement et d'hygiène 2000-2020

20. In Asia, 50% of urban areas (2.4 billion people) are located in low-altitude coastal areas and so are under the influence of sea level rise according to the UN WATER 2020 report Water and climate change

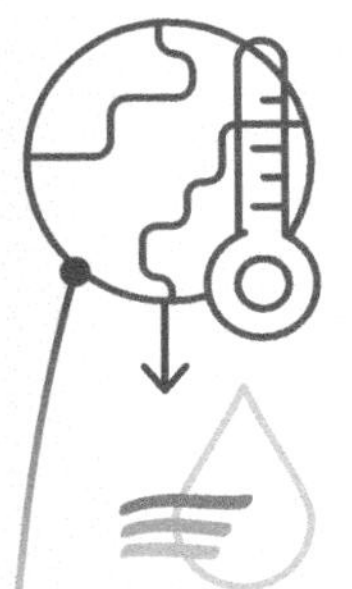

3.2 The impacts of global changes on water and sanitation services

Climate change and the **rise of the average temperature** result **in long-term shifts in climate trends**: global warming of the atmosphere and oceans, a faster rise in sea levels and changes in rainfall patterns depending on the region. Climate change also causes **the frequency and intensity of extreme weather events to increase**[21]: torrential rains, cyclones, droughts, heatwaves. These events may directly impact water and sanitation services in terms of **quality and availability of water resources**, and alter their **operation and viability** over time.

In France, the impacts of climate change on water and sanitation infrastructures appear as an **increased risk of flooding, increased sewer overflows, enhanced challenges to protect river quality**, as well as infrastructure damage resulting from the shrinkage and swelling of clay soils. Another possible impact is the temporary or definitive **flooding** of certain areas caused by the increased coastal erosion that goes with the **rise in sea levels**[22].

On a global scale, pressure on water and sanitation services is also due to these heightened climate risks and the **rise in water demand** brought about by the combined effects of **economic development, shifts in consumer habits, higher temperatures** and **population growth**: in 2019 the UN estimated that world population would 9.7 billion people in 2050. With these new challenges, water and sanitation services must anticipate local risks by adopting an adaptation approach. All stakeholders globally, including water and wastewater services, still have lots to do to curb their greenhouse gas emissions and thus limit the extent of the risks they face.

ADAPTATION AND MITIGATION IN THE WATER AND SANITATION SECTOR: WHAT ARE THE LINKS?

To this day, water is still not sufficiently integrated into climate policies. In the water sector **priority has been given to adaptation** rather than **climate change mitigation**.

Yet the adaptation and mitigation approach are complementary. Actions to reduce GHG emissions offer co-benefits in terms of adaptation (e.g. reducing water losses can help to adapt to increased water scarcity and reduce the use of fossil energy for pumping) and are even necessary so as to avoid irrelevant adaptation measures. Conversely, adaptation measures can also offer co-benefits for mitigation as detailed in Chapter 4, for example green infrastructure to reduce flooding can also reduce the volume of water treated at wastewater treatment plants for combined sewer systems. When irrelevant adaptation measures are taken, for instance regarding installations, they may create lock-in effects that will make it difficult, if not downright impossible, to later rehabilitate ecosystems and infrastructures and may have economic, social and environmental consequences, for example higher costs, while other emergencies crop up. Investments made by water services must consider the mitigation/adaptation links to better address new constraints associated with climate change, such as: decreases in the allowable loads discharged to waterways, combined sewer overflows as heavy rainfalls increase, etc. Not considering these links may cause GHG emissions to increase thus adding to climate change and less resiliency to its effects.

> In the water sector priority has been given to adaptation rather than climate change mitigation.

21. IPCC, AR5 Synthesis Report: Climate Change 2014

22. ASTEE, Guide Eau Déchets et Changement climatique 2019

Large amounts of energy are required to capture, treat and transport water for human needs, particularly in urban areas. Producing electrical energy requires, directly or indirectly, vast amounts of water. This interdependence between the two resources is called the **Water-Energy Nexus**[23]. It can include carbon as well if taking into account these two sectors' dependency on fossil fuels (coal, oil, etc.) and their GHG emissions. Fossil fuels (coal, oil, gas), the main causes of global warming, still accounted for 80% of final energy consumption worldwide in 2019, as in 2009, according to the most recent
REN21 report[24] (June 2021). The demand for energy is expected to remain on the increase in coming years, meaning that the energy sector will likely put water resources under heavy pressure. Conversely, the water sector is likely to consume more and more energy, either to draw up on alternative sources of water, to carry out more advanced treatment of water, or to treat larger volumes of water, as a result of climate change and global changes.

3.3 Regulatory and political framework

Reducing the GHG emissions of the water and sanitation services is part of the broader battle against climate change and its impacts, as monitored by the **Intergovernmental Panel on Climate Change** (IPCC).

The **Paris Agreement**[25] set as target to "**keep global warming well under 2°C compared to pre-industrial levels**" and to try to limit the rise to 1.5°C by the end of the 21st century. To reach these targets, nations must submit their plans for adaptation and mitigation, known as **nationally determined contributions** (NDCs) which concern all sectors including that of water. Countries/Parties, as well as local authorities, may base their mitigation actions by water and sanitation services on **Agenda 2030**[26] which sets 17 **Sustainable Development Goals** (SDGs). Two are particularly relevant: **SGD 6 — Ensure availability and sustainable management of water and sanitation for all"** - and **SDG 13 - "Take urgent action to combat climate change and its impacts".**

IN FRANCE:

Mitigation action by the water and sanitation sector is part of the broader national approach for an energy transition (law n° 2015-992)[27]. The goal is to make final energy consumption decrease by 50% by 2050 compared to 2012 and to divide GHG emissions by four. These goals concern all sectors and actors. The law known as Grenelle II makes it mandatory for companies employing more than 500 people, organizations employing more than 250 people and cities with more than 50,000 inhabitants to carry out carbon footprint assessments; in addition the local authorities concerned must draw up a Climate Air quality Energy plan at the level of their territory. The plan is to extend these requirements to smaller organizations in the near future.

---

23. Water and Energy, Threats and Opportunities Olsson, 2015

24. REN21 Renewable Global Status Report, 2021

25. The Paris Agreement was adopted in 2015 at the Conference of Parties (COP) as part of the United Nations Framework Convention on Climate Change (UNFCCC)

26. United Nations Program for Sustainable Development or Agenda 2030 (2015)

27. Loi relative à la transition énergétique pour la croissance verte (2015)

28. (LTECv) (Art 173-vI).

Investors are expected to divest from fossil fuels and the most emission-heavy sectors, and instead focus on renewable energy and sectors that are driving the transition, as the law on energy transition and for green growth states[28] .

Many EU member-states, France included, have signed the **European Green Deal**, a set of policy initiatives aiming to reach **carbon neutrality[29] by 2050**. France has translated this goal into a roadmap known as its National Low-Carbon Strategy (**Stratégie Nationale Bas Carbone**, SNBC). In the UK, the water industry has defined a Roadmap to Carbon neutrality with a Net Zero target by 2030[30] .

Legislation pertaining to the reuse and conversion of resources is not yet stable, because certain ways of reusing wastewater or sewage sludge still lack acceptability. The European circular economy action plan presented by the European Commission supports the development of the reuse and recycling of water in the European Union. Worldwide, the legal framework is that of the **Compendium of Water Regulatory[31]** which regulates water usage.

Therefore, it is clear from just the NDCs and Paris Agreement that drivers exist for GHG mitigation in the water sector. Furthermore, as part of the UNFCC Race to Zero initiative[32] , Global Water Intelligence has compiled a list of water utilities with net zero goals: The Net Zero Utilities Observatory[33] , which tracks the timeframe for their targets (e.g., 2030, 2050); type of target; current greenhouse gas emissions; whether they have joined the UNFCCC Race to Zero campaign; and the city and population served.

4. HOW ARE GHG EMISSIONS ASSESSED?

Many carbon accounting systems are available at international level for institutions and companies. The GHG Protocol seems to be the most widely used. In France, an NGO (Association Bilan Carbone) has completed and improved a carbon footprint tool, Bilan Carbone, initially developed by the ADEME (France's energy transition agency). However these carbon accounting systems are not specific to the water sector. In the French case, the **ADEME and ASTEE** (Scientific and Technical Water and Waste Organization) have produced a sectorial guide, adapting the Association Bilan Carbone's method to the water and sanitation sector[34]. Other countries have developed similar guides which sometimes differ regarding the emissions that should be accounted for in the assessment methods, particularly for items that fall under "Scope 3" (refer to chapter 1.2). These indirect emissions result from the activity in question but the definition of their scope may vary depending on national approaches.

These indirect emissions result from the activity in question but the definition of their scope may vary depending on national approaches.

---

29. Carbon neutrality is defined as a state of balance between the carbon dioxide emitted into the atmosphere and the carbon dioxide removed from the atmosphere.

30. https://www.water.org.uk/routemap2030/

31. UN WATER Compendium of water regulatory (2015)

32. https://unfccc.int/climate-action/race-to-zero-campaign

33. https://www.globalwaterintel.com/water-without-carbon

34. ADEME ASTEE guide méthodologique des émissions de gaz à effet de serre des services d'eau et d'assainissement mis à jour en 2018

Another limitation of current GHG accounting methods for Scope 1 emissions of wastewater N_2O and CH_4 process emissions is that they fail to account for the site-specific conditions that actually dictate the extent of emissions. For N_2O for example, based upon the IPCC guidance, a single global emission factor is proposed for Tier 1 and a country-level emission factor is proposed at the Tier 2 level. Tier 3 is most accurate because it is based on site-specific measurements and actual N_2O emissions at a particular site. However, Tier 3 is not always feasible and water utilities need to report fairly accurate emissions in the meantime. The emission factor approach for Tiers 1 and 2 only considers the influent nitrogen load, which does not dictate to what extent a site will have N_2O emissions. This is discussed in further detail in the IWA publication Quantification and Modelling of Fugitive Greenhouse Gas Emissions from Urban Water Systems (Ye *et al.*, 2022).

The IWA and GIZ have developed the tool "Energy Performance and Carbon Emissions Assessment and Monitoring" (ECAM) as part of the Water and Wastewater Companies for Climate Mitigation (WaCCliM[35]) program. This tool supports operators, mostly in developing and emerging countries in their GHG accounting and reporting at national level, but discussions are still ongoing to standardize the approach to scope 3 and avoided emissions.

Avoided emissions may be taken into account if the by-products are reused by a third party as a substitute for a "product" or if the energy is produced by the plant and used by a third party. The GHG required by the third party to produce this product, or to produce this amount of energy were not emitted, so they are deemed "avoided emissions".

---

35. Project Water and Wastewater Companies for Climate Mitigation, on behalf of the German Federal Ministry for environment, Nature Conservation and Nuclear Safety (BMU).

Energy used worldwide
by water and sanitation service providers

While energy utilities use water for producing energy, water utilities use energy for providing water, one example of the so-called water & energy nexus. The International Energy Agency estimates that in 2014 the water sector used up 1400 TWh worldwide , 60% of which was consumed as electricity and 40% as thermal energy, mainly to pump groundwater for irrigation. The water sector accounted for 4% of global electricity consumption in 2014, excluding the energy related to the use of water by end users. Of the electricity consumed for water, around 40% is used for **water extraction**, 25% for wastewater treatment and 20% for water distribution. Another 5% is used up for desalination of seawater. There is however a big difference between high and low income countries. In high-income countries, wastewater treatment makes up the lion's share of electricity consumption (42%). This proportion is lower in low-income countries because a large portion of wastewater is neither collected nor treated. As a result of global warming and increased water stress in certain parts of the world, energy consumption in the water sector is projected to double by 2040 in the IEA scenario. The reason lies in part in the growth of desalination and water transfers but also in population growth and urbanization which will increase energy consumption of treatment plants.

...60% of which was consumed as electricity and 40% as thermal energy, mainly to pump groundwater for irrigation.

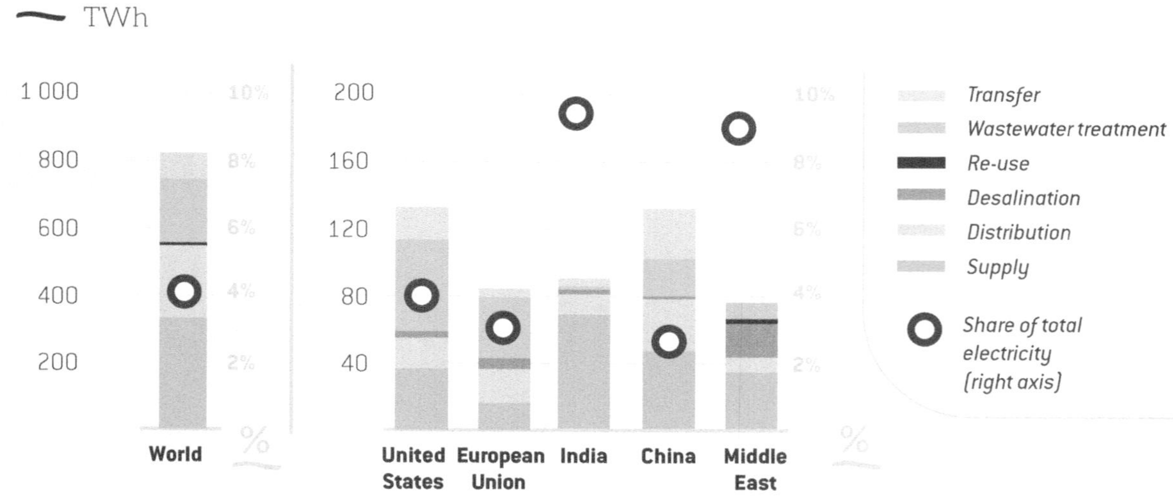

The water sector accounted for 4% of global electricity consumption in 2014

When looking at the detail by geographical region, it appears that the United States is the world's largest electricity consumer for the water sector (40%), while China accounts for 15% of global electricity needs in the water sector and the Middle East, where desalination is a major activity, for 9%. A large share of India's electricity consumption by the water sector goes to groundwater extraction for irrigation.

The carbon footprint associated with the energy consumption of water and sanitation services depends very much on the local energy mix. The electricity and energy used are often derived from fossil fuels such as **coal, oil or natural gas,** which emit large amounts of greenhouse gases per unit of electricity generated.

1. WATER ABSTRACTION

Globally, water extraction is estimated to consume over 310 TWh1[36] of electricity per year. The amount of energy required depends on the source (**groundwater pumping is roughly 7 times more energy intensive** than surface water abstraction) and on the distance and elevation that the water must travel before reaching the storage and treatment facility. Globally, surface water accounts for around 2/3 of all water withdrawals
and groundwater for about 1/3. In some countries, the water resources are not sufficient to meet demand. Consequently, some countries have embarked on large-scale water transfer projects. In low-income countries, surface water can be heavily polluted. Groundwater should be preferred for the supply of drinking water, because especially at depth, it is less likely to be contaminated and is less affected by strong variations like in the flow of surface water. However, this involves protecting abstraction sites (establishing a protection perimeter or forbidding polluting activities near capture sites, etc.).

---

36. International Energy Agency World Energy Outlook Excerpt Water energy Nexus, 2016

Unfortunately, too many groundwater resources are not properly protected, resulting in a decrease of their piezometric level, and contamination, often by agricultural pollutants. Water consumption in India and the Middle East is sourced almost exclusively from underground resources, resulting in local overexploitation, mainly for agricultural purposes. In many regions this has caused lower groundwater tables, which in turn causes high electricity needs. In other places, decontamination of polluted groundwater is required.

Drinking water is often extracted from rivers, which implies being particularly vigilant regarding variation in flows and surface water contamination. China and the United States consume almost exclusively surface water. France's policy is to protect groundwater and use it as a preferred resources for drinking water supplies.

Finally, one alternative mobilized by many coastal municipalities suffering from a water deficit is sea water desalination, an energy-intensive technology but nonetheless very much used in particular around the Mediterranean coast. Desalination can be an efficient alternative to long-distance water transfers.

2. WATER STORAGE (FOR DRINKING WATER)

Reservoirs are one way of addressing the challenge of flow irregularity, particularly low-water flow, or guaranteeing supplies. Reservoirs used solely for drinking water are generally installed on small rivers supplied by good-quality water, and are relatively small. They have to be protected well, like catchments from underground water.

At the same time, larger multipurpose reservoirs can produce hydropower energy and regulate flows for uses such as water supply, irrigation, flood control, environmental management and pollution control. The reservoirs themselves can provide for activities such as navigation, fisheries, recreation and, most recently, as a space available for other forms of energy generation such as floating solar PV. The challenge is in the optimisation of a reservoir's multiple services, while recognising that priorities may change over time. None of hydropower's multipurpose benefits will be realised unless projects are built in the right place and in the right way. Not all negative impacts can be avoided, and mitigation and compensation actions can be as important as the project itself.

Far greater accountability and transparency can be achieved today on reservoir multiple benefits management. Tools exist for measuring sustainability performance in the planning, preparation, implementation and operation stages, such as the Hydropower Sustainability Guidelines developed by the IHA (International Hydropower Association) with the concerned stakeholders.

3. WATER CONVEYANCE

It is unfortunately frequent to have to resort to long-distance water transfers (China, California, Morocco, Tunisia, South Africa, etc.) after abandoning nearby resources for reasons of quantity (water stress) or quality (water pollution). The drawdown of groundwater[37] after overexploitation of the resource can also lead to water transfers. These represent 70 TWh annually on a global scale, or more than the treatment of drinking water (65 TWh).

---

37. Decrease to piezometric zero (maximum) water table levels caused by pumping or natural or accidental drainage of the groundwater.

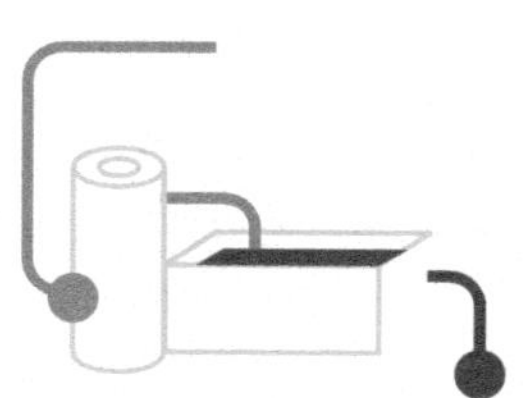

4. WATER TREATMENT

In water treatment plants pumping and treatment use up the most energy. **Water treatment requires 65TWh[38] of electricity globally**, of which 80 to 85% are used for pumping. Extracting underground water uses more than extracting surface water, but the energy requirements for treating it are lower because it is usually less polluted. The problem of diffuse pollution in water is an important factor in GHG emissions during treatment because it means that more complex, hence more energy-intensive, treatment is required. These treatments concern both macro pollutants such as nitrates (denitrification) and micropollutants such as pesticides and pharmaceutical residues, which require removal technologies such asactivated carbon, reverse osmosis or ozonation. With the rise in temperature, the concentration of pollutants in the various water resources is expected to increase and make their treatment more difficult.

5. WATER DISTRIBUTION

Bringing drinking water under pressure from the water treatment plant to the consumer requires a lot of energy. Water distribution on a global scale uses **about 180 TWh[39]** but the amount of energy used for this stage varies greatly depending on the region.

Water losses (leaks, stealing of water, inadequate measurement, etc.) substantially impact energy consumption, particularly in countries where the production of drinking water consumes large amounts of energy. Water losses in public supply are estimated at 12% in the United States, 19% in China, 24% in the European Union and 48% in India.

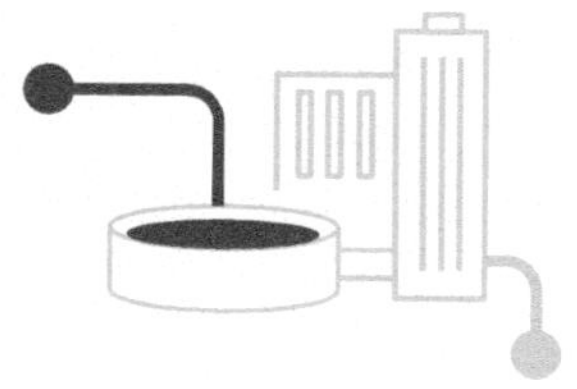

6. TREATMENT OF WASTEWATER

On a global scale, **the treatment of wastewater represents about 200TWh[40]**. In high-income countries, wastewater treatment is the primary energy consumer of the water sector. For some municipalities, the energy consumed by water and wastewater utilities can account for 30-50% of their energy bill (United States Government Accountability Office, 2011). In emerging economies, wastewater treatment represents a large potential demand for energy.

Five factors affect the quantity of energy required for treating wastewater: **the amount of wastewater collected and treated, the infiltration and inflow** (water from underground and rain sources) in the wastewater network, the level of treatment to meet **discharge requirements, the type and concentration of pollutants in the wastewater, and the energy efficiency of the process.**

38, 39, 40. International Energy Agency, World Energy Outlook, Excerpt Water Energy nexus 2016

The level of treatment varies depending on the world region. In some countries in Asia and Africa, primary treatment is predominant, while in OECD (Organization for Economic Co-operation and Development) countries, secondary is standard and some countries also use tertiary treatments. About **50 % of the energy required for waste water treatment** is used for **secondary treatment, in particular for aeration**. Pumping and management of sewage sludge are also heavy users (respectively 16% and 15%) even though recycling sludge could produce energy (estimated to **6TWh worldwide**).

7. WATER END USE

A review of global studies indicates up to 12.6% of total national primary energy use can be influenced by water, when water-related energy of water end users, is included as well as the more commonly considered energy use by water utilities, are all included (Kenway *et al.* 2019 - see **figure 1, page 9**). Water heating for residential, commercial, and industrial purposes is the dominant fraction accounting for over 90% of the effect. Water and wastewater utilities use 0.4-2.3% of primary energy or 0.6-6.2% of regional electricity, mostly for water pumping.

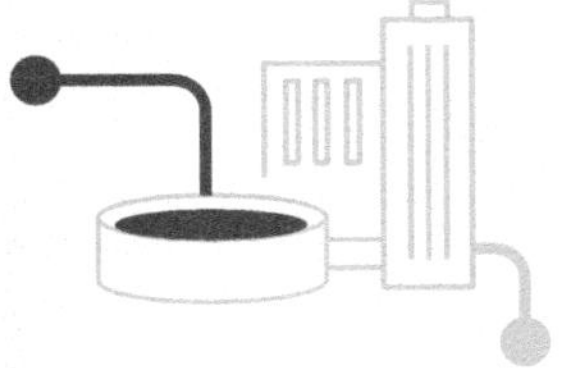

8. DESALINATION AND WATER REUSE

According to the World Bank[41] , in 2018, 18,426 desalination plants were reported to be in operation in over 150 countries, producing 87 million cubic meters of clean water each day and supplying over 300 million people. There are two main desalination methods, thermal and membrane, which can be combined as a hybrid. Thermal desalination is a process of boiling and evaporating salt water and condensing the resulting vapor. Membrane methods adapt the natural process of osmosis, and reverse osmosis (RO) is the most commonly used form due to its lower energy consumption. Even though desalination represented only 0.7% of water needs in 2016, it represented 25% of the energy consumed by the water sector and 5% of the global electricity. However, the significant progress made has led to reduce the energy intensity by a factor of 10 from 20-30 kWh/m^3 in 1970 to about 3 kWh/m^3 in 2018. Less than 15% of the energy is provided in the form of electricity, the rest by fossil fuels, with natural gas being the preferred fuel for thermal desalination.

Water re-use describes the use of discharged wastewater as a source of freshwater. In general, a distinction is made between potable and non-potable re-use with non-potable re-use employed mainly for irrigation. As drinking water needs to meet higher quality standards, the process of re-using water for potable purposes requires more energy than non-potable purposes.

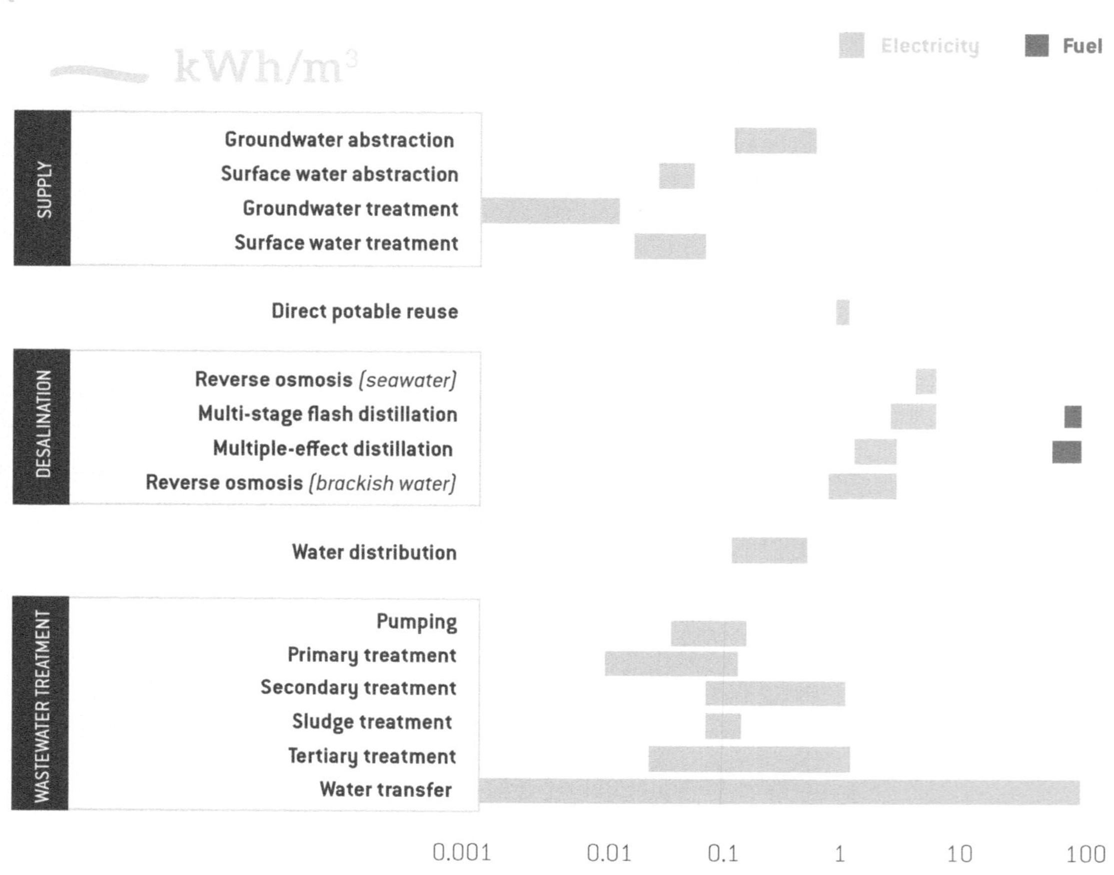

Seawater desalination and wastewater treatment are the most energy-intensive processes in the water sector

STAGES- WATER AND SANITATION SERVICES	ELECTRICITY CONSUMPTION IN KWH/M³ FRANCE [42]	ELECTRICITY CONSUMPTION IN KWH/M³ WORLD [43]	GLOBAL ELECTRICITY CONSUMPTION [44]	EQUIVALENT IN CARBON TONS WITH EUROPEAN ENERGY MIX [45]
TOTAL WATER SERVICES			820 TWh	204Mt CO_2 eq
DRINKING WATER TREATMENT	Global 0.5 kWh/m³ Conventional treatment 0,05-0,15, membrane 0,1-0,2, advanced 0,25-0,7	0,001-0,09 kwh/m³ (0,001-0,02 groundwater) (0,03-0,09 surface water)	65 TWh	16 Mt CO_2 eq
EXTRACTION			310 TWh	77 Mt CO_2 eq
REUSE	0,025-1,05 kWh/m³	0,9-1 kWh/m³		
DESALINATION OF SEAWATER	3-8 kWh/m³	1-9 kWh/m³		
DESALINATION OF BRACKISH WATER	0,6-1,5 kWh/m³	0,9-3 kWh/m³		
WATER TRANSFERS		0,001-100kWh/m³	70 TWh	17 Mt CO_2 eq
DISTRIBUTION	0,1 kWh/m³	0,2-0,7kWh/m³	180 TWh	45Mt CO_2 eq
WASTEWATER COLLECTION/ TRANSPORT	0,06 kWh/m³	0,04-0,3kWh/m³		
PRIMARY TREATMENT		0,02-0,2kWh/m³		
SECONDARY TREATMENT	0,2 kWh/m³	0,08-1kWh/m³		
TERTIARY TREATMENT		0,03-1kWh/m³		
TOTAL WASTEWATER TREATMENT			200 TWh (sludge recycling -6TWh)	50Mt CO_2 eq
THERMAL DRYING OF SLUDGE		0,08-1kWh/m³		

42.Figures drawn from Guide AMORCE Les services d'eau et d'assainissement et changement climatique: les leviers d'atténuation source Veolia Eau

43.Figures from a diagram in International Energy Agency, World Energy Outlook, Excerpt Water Energy nexus 2016 Sources: EPRI (2002); Pabi, et al. (2013); Jones and Sowby (2014); Plappally and Lienhard V (2012); Spooner (2014); Li, et al (2016); Japan Water Research Center (n.d.); (Choi, 2015); Miller, et al. (2013); Singh, et al. (2012); Noyola, et al. (2012); Liu (2012); DWA-Leistungsvergleich (n.d.); Caffoor (2008); World Bank Group, (2015); Fillmore, et al. (2011); Brandt, et al. (2010); IEA analysis.

44.Figures taken from International Energy Agency, World Energy Outlook, Excerpt Water Energy nexus 2016

45. Calculated using the European carbon factor 2020 249 Kg CO_2/MWh

THE FRENCH CASE: GHG emissions from energy consumption by water and sanitation services

According to the ADEME[46] energy consumption for water services in France produces an annual output of **4.2 kg CO_2 equivalent/inhabitant for drinking water and 5.9 kg CO_2 equivalent /inhabitant for collective sanitation.**

In France, between **30 and 55% of energy consumption for public water and sanitation concerns sanitation**[47] **and 90%** of this energy expenditure is associated with treatment of urban wastewater (and the treatment of by-products for their reuse) which is the most energy-consuming item. Aerating wastewater in plants has high energy consumption averaging 0.2kWh/m³.

46. ADEME et ASTEE Guide des émissions de gaz à effets de serre des services d'eau et d'assainissement, réédité 2018

47. A.E Stricker *et al.*, Consommation énergétique des filières intensives de traitement des eaux résiduaires urbaines, 2018

3 Reducing the GHG emissions *of* water and sanitation services: what enablers and solutions?

This chapter provides an overview of enablers and solutions to reduce the greenhouse gas emissions of water and sanitation services, organised in three categories:

- **1/** reducing water and energy consumption through lean and efficient approaches,
- **2/** embracing circular economy to produce energy and valuable products, and
- **3/** planning to reduce GHG emissions through strategic decisions that underly the everyday operations.

As noted in Chapter 1, the first step to reducing GHG emissions is to properly assess and monitor these emissions from year to year, recognising that some emissions are operational and others associated to infrastructure investments. This enables for a targeted and structured approach to planning and implementing actions. For example, the City of Chiang Mai in Thailand (see **Project #3-FWP**) used the afore-mentioned Energy Performance and Carbon Emissions Assessment and Monitoring (ECAM) tool to diagnose GHG emissions at different stages of the sanitation service. It was determined that a 12% reduction in GHG emissions could be achieved through two main levers: energy optimization of pumps and repair of leaks in the wastewater system.

1.   LEAN AND EFFICIENT APPROACHES

This section focuses on technical measures or approaches that lead to reducing the amount of water and energy consumed as a means to reduce the GHG emissions.

1.1   Network maintenance to reduce water loss and infiltration

Sound asset management enables to reduce the risk of leaks and pipe bursts in drinking water networks, as well as clear water intrusions in the sewer networks. It is a key element to save energy and reduce greenhouse gas emissions, as the volumes of water treated unnecessarily result in wasted energy. Water loss and sewer dilution indicators are used in asset management to quantify, locate and repair damaged pipes. Beyond this curative approach, asset management methods are also available to anticipate future leaks and make targeted maintenance investments prior to the pipes being damaged.

For sewerage networks, the issue is the permanent infiltration of parasitic clear water from groundwater but also rainwater entering the network after a rain event (see **project #3-FWP**). The consequences are as follows:

> sewer overflows that cause GHG emissions (methane and nitrous oxide) in the receiving water bodies,

> wear and tear on the infrastructure that cause premature investments and GHG emissions associated to construction material and equipment manufacturing, and

> greater energy consumption for pumping and treatment, that cause GHG emissions proportional to the country's energy mix, and to the greater capital infrastructure

Several innovative approaches are being developed to support utilities taking action. Basic action to be undertaken is to compare expected and actual volumes of treated wastewater to assess the level of infiltration (from groundwater) and inflow (from rainwater). If these levels are high, new technologies could valuably help locate the origin of the infiltration and inflow. For example, Suez is developing an innovative mobile device with embedded quality sensors (T°C, conductivity, pH and redox), collecting data at high frequency, to monitor the wastewater quality along a pipeline. An abnormal change in the monitored signals could be the result of two volumes of water of different quality mixing. A dedicated algorithm has been developed to automatically identify disorders in sewers, to support targeted and timely repairs by the utility.

For drinking water networks, leaks and pipe bursts are responsible for greater energy consumption for pumping and treatment, that cause GHG emissions proportional to the country's energy mix. However, the main drive for taking action is often conserving water and reducing water abstraction requirements. Digital tools are available to support drinking water utilities. For example, the SmartWater system implemented by SUEZ, or HpO proposed by Alteréo, are being used for proactive reduction of water loss. These tools make it possible to remotely monitor the flow rates and pressure of pipes. Also, satellite observation technology could help with detection of leakages. PVK operating Prague water network cooperates with UTILIS company and satellite observations helped to reveal 50 leakage points in 2020 . Some of these tools may also be used to inform individuals about their consumption and, when needed, warn them in the event of a leak.

48.https://www.pvk.cz/aktuality/satelit-odhalil-pres-dve-ste-potencialnich-uniku-pitne-vody-patraci-potvrdili-padesat-lokalit/

1.2　Efficiency of services

The efficiency of water and wastewater services can be improved in regards to capital investments or operational measures, as presented below:

Capital investment related efficiency:

- Improving the **efficiency of motors and pumps** (see **project #3-FWP**).
- Improving the **design and operation of pumping stations** to reduce the energy requirements per m³ of water transported. Regulating the flow rate through a series of pumps or a variable speed drive enables to move water from one point to the next close to the maximum efficiency point of the pumps (see **project #11-ASTEE**).
- **Adapting the pressure in the water pipes and regulating the water storage levels** to reduce the energy consumption of water distribution, as well as reduce the risk of new leaks and the volume lost through existing leaks.

Operational measures related efficiency:

- **Optimizing drinking water treatment:** The treatment of water depends on the quality of the water abstracted. Advanced treatments are the most energy consuming, therefore investing to preserve the source quality should be the priority for long term planning. In the short term, energy can be saved by optimizing the process itself or selecting low-energy processes during the design phase. The volume and type of reagents necessary to the treatment may be reduced or adapted to reduce their associated GHG emissions. The energy and GHG efficiency plan of Eau de Paris can be an inspiring example of this approach (see **project #5-ASTEE**)
- **Optimizing wastewater treatment:** Planning future plant with modular infrastructures that can be adapted to the load and flow to be treated, is a way to avoid oversizing and underloading, which result in an overuse of energy per volume of water treated. An energy audit of activated sludge plants may reveal a potential to reduce energy consumption in particular on the aeration, pumping, and biosolids handling. Energy may also be saved by not treating the wastewater more than it needs to be treated to meet the discharge requirements, in countries where these are set based on the natural capacity of the waterbodies to absorb pollution. Beyond energy, the wastewater treatment process may be improved in the following way:
    > fine-tuning the aeration and flow balancing of the process to reduce nitrous oxide emissions. Innovative control systems are being developed to support utilities with this approach
    > Reducing the reagents needed for treatment. For example, the Grenoble metropolitan area's WWTP (Grenoble Alpes Métropole) has removed the need for reagents in their lamellar primary sedimentation tank, through physical hydraulic improvements (see **project #6-ASTEE**)
    > When retrofitting or building new infrastructure, considering technologies such as mainstream anaerobic treatment[49] or membrane aerated biofilm reactor (MaBR)[50] can lead to reduced nitrous oxide emissions.
    > Biosolids from activated sludge systems are non-self-heating and require energy (gas or fioul) when incinerated. This energy consumption can be reduced using digital algorithms (see **project #10-ASTEE**).

49. https://www.cranfield.ac.uk/research-projects/demonstrating-next-generation-circular-water-solutions

50. 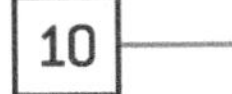2020 Report from Danish EPA

- **Monitoring of wastewater networks for industrial discharges:** Wastewater network monitoring aims to reduce non-compliant industrial wastewater discharges. These discharges could bring excessive amounts of nutrients, fats or toxic chemical compounds and thus increase GHG emissions of the wastewater treatment plant. These discharges may be monitored through sampling campaign, on-line measurements of basic parameters and colour cameras.

- **Reducing the transport of personnel, supplies and by- products:** even before considering the transition to renewable energy transport, it is often possible to significantly reduce the distances and number of trips needed to ensure a quality service. For example, Grenoble metropolitan area's WWTP (Grenoble Alpes Métropole) is reducing transport of by-products by improving the screenings residuals compacting and investigating the injection of the grease collected into the on-site digestor in place of trucking it to a remote facility (see **project #7-ASTEE**). Transport may also be reduced through preventive maintenance reducing vehicles usage for emergency repairs, and though using efficient route planning, and possible use of alternative fuels.

1.3 Reducing water and water-related energy consumption by end users

GHG emissions of the urban water cycle are most important at the end users. Inspiring and incentivizing industries and individual households to reduce their water consumption and their energy requirements for water heating are very impactful measures to reduce GHG emissions. These measures rely mostly on information and awareness approaches as discussed in section 3.3, but also on physical measures such as low-flow fixtures and appliances, or improving the energy efficiency of water heating. Individual meters may enable to inform end users of their consumption and provide incentives to reduce it. The utility is usually not a direct implementer of these measures, but has an influencing role to make the change happen (see **project #2-FWP**).

An example of a measure taken by a utility is the de-carbonatation of the drinking water supply by SEDIF, in the greater Paris area (see **project #3-ASTEE**). By softening the water supply, SEDIF reduces the scaling on water-heating appliances which become more efficient and increase their life expectancies. The life cycle analysis of this initiative already shows a positive GHG reduction, without accounting for the increased consumption of tap water for drinking rather than bottled water. The latter impact is hard to measure, but estimated to significantly reduce the emissions related to the whole of the urban water cycle.

1.4 Low-impact wastewater treatment

A combination of green and grey infrastructures can support water and sanitation services that are cost effective and very low carbon. For example, the City of Notse in Togo has implemented a solar panel based water supply combined with a latrine and wetland treatment system to provide the inhabitants the basic services they needed (see **Project #4-FWP**). These projects tend to be frequent in emerging economies, and inspire more and more developed cities to consider decentralised approaches in their peripheries in place of expanding centralised treatment capacities (see **project #8-ASTEE**). Considering extensive and carbon effective solutions should be included in the investment plans and design process of drinking and wastewater services, especially in rural areas.

1.5   Alternative rainwater management

Alternative rainwater management is about retaining and infiltrating rainwater where it falls to avoid the energy requirements associated to its transport and treatment when discharged either to a combined sewer, or a separative rainwater network. These measures combine Nature based Solutions (NbS) such as green roofs, retention ponds, rain gardens, or swales and reducing the percentage on non-permeable urban surfaces, as presented in Chapter 4 (see **project #12-FWP**). Beyond the energy savings, these alternative rainwater management methods also provide a small amount of carbon sequestration in vegetation and soils, which is not well quantified at this stage. Separate rainwater treatment systems may provide lower carbon solutions – either through conventional treatment unit operations (physical, chemical, biological) or through nature based treatment wetland solutions for stormwater which can provide wider public benefits. For example, a Florida stormwater treatment wetland system is also a tourist attraction with a trip advisor rating.[51]

1.6   Restoring and preserving the quality and quantity of water resources

Investing in the protection of the source water quality reduces the need for treatment and its associated energy requirements and GHG emissions (see **project #13-FWP**). Preserving the quantitative balance of a resource, removes the need for long distance water transfers, or energy intensive treatment of a local water supply of lower quality. Often these approaches involve restoring catchments which may have co-benefits of restoring soil carbon and improving soil system health. They can therefore provide wider carbon benefits albeit not accounted for by Companies in reported emissions.

This approach requires long term planning and partnerships across institutions and geographic zones, but can provide both a reduced GHG footprint of water services, and an increased resilience to the impacts of climate change, as highlighted in Chapter 4.

---

51. https://www.tripadvisor.co.uk/Attraction_Review-g34467-d6366892-Reviews-Freedom_Park-Naples_Florida.html

2. CIRCULAR ECONOMY

This section focuses on technical measures or approaches that lead to producing renewable energy within water and sanitation services, or to recycling materials from wastewater as a means to reduce GHG emissions. Circular economy is a disruption to the linear economy model (extract, manufacture, consume, throw away) for a circular economic model (produce and consume what is needed, (re)use all by-products and regenerate natural systems). In the water sector, we can apply the so-called 5R rule of the circular economy: Reduce, Reuse, Recycle, Recover and Restore[52].

The "Reduce" and "Restore" component are discussed in the previous section. However, it is useful to keep in mind here, that reducing the needs is the first step of a circular economy approach, and that restoring the environment is essential to projects' sustainability[53].

2.1 Reusing water, nutrients and materials

Greywater, wastewater or stormwater can be reused at different levels of treatment, e.g. as non-potable water for irrigation, amenity water and public cleaning. This approach is both an adaptation and mitigation measure, as it allows to deal with water scarcity, and may reduce the energy required for treatment to meet the requirements for these usages rather than that for discharge to a water body. Local decentralised treatment also saves emissions related to operation and maintenance of distribution networks. Moreover, greywater local treatment requires less energy than centralised wastewater treatment. However, these solutions need to be assessed through a life cycle analysis (LCA) that account for all energy and non-energy related emissions, in order to ensure that they actually contribute to reducing global GHG emissions[54].

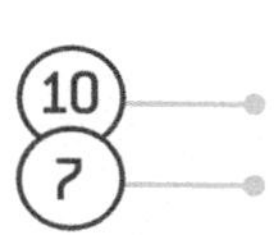

An example of treated wastewater reuse is in eThekwini, where recycled water is supplied to local industries (see **Project #10-FWP**), or at smaller scale in Gaza where greywater recycling units enable local reuse of non-potable water by the inhabitants (see **Project #7-FWP**). The project aims to equip a hundred households with grey water treatment systems for use in irrigation of a tree farm.

When recycling treated wastewater in agriculture some nutrients can also be recycled as fertiliser. For example, in Clermont Ferrand, France, urban wastewater mixed with that from a local sugar factory started to be reused in 1996[55] to irrigate about 750 ha of cereal crops in the Limagne Noire plains. This land lacked underground resources and water transfers from the Allier River were not feasible. Another example is the As Samra plant run by Suez that treats water from Greater Amman in Jordan and recycles the wastewater to irrigate crops (see **Project #9-FWP**).

Wastewater or rainwater can also be used to provide drinking water. This is most often done as indirect reuse by reinjection into groundwater like in Windhoek, Namibia (see **Project #8-FWP**) or into surface water bodies after extensive treatment like in Singapore. In this case, water reuse is very energy intensive and is implemented as a climate adaptation measure.

---

52. IWA 2016, Water utility Pathways in a Circular Economy.

53. UKWIR 2021, What does a circular economy water industry look like?

54. References on LCA and ecoefficency methods applied to the water sector work. e.g.
Farago *et al.* 2019, An eco-efficiency evaluation of community-scale rainwater and stormwater harvesting in Aarhus, Denmark (**https://www.sciencedirect.com/science/article/abs/pii/S0959652619302902?via%3Dihub**)
Farago *et al.* 2021, From wastewater treatment to water resource recovery: Environmental and economic impacts of full-scale implementation (**https://www.sciencedirect.com/science/article/pii/S0043135421007508?via%3Dihub**)

55. by the Association Syndicale Autorisée (Asa) Limagne Noire

In such projects, the societal component is a critical success factor to ensure the acceptability of the wastewater reuse, in particular in the case of the "toilet to tap" concept. The acceptability of this type of solution varies according to the regions of the world. In some Asian and African countries, wastewater is commonly used to fertilize and irrigate crops.

Nutrients may be recycled through land application of the biosolids produced at wastewater treatment plants. For example, the greater Paris recycles over 60% of their biosolids in agriculture. At the As Samra plant in Jordan, the sludge is also recycled in the form of granules for reuse as fertilizer or fuel (see **Project #9-FWP**). Land application of biosolids has an agronomic interest through the improvement of soil quality (better carbon content and water absorption), especially when the sludge is digested or composted. However, the composting of sludge is still limited in France due to social acceptability and regulatory obstacles regarding the mixing of different types of waste streams. In addition, land application of biosolids is currently evaluated in regards to microplastics and per and polyfluoroalkyl (PFAS) substances contained in the influent wastewater.

Innovative processes are also being implemented in few places to recover nutrients from biosolids dewatering waste streams, therefore avoiding the energy demand and potential GHG impact from emissions of nitrous oxide when removing these nutrients as they are sent back to the head of the treatment plant, or from ashes when biosolids are used to produce heat. The use of recovered nitrogen and phosphorus in agriculture could avoid GHG emissions by replacing conventional fertilizers. Nitrogen fertilizers are typically produced using a chemical process that requires a lot of energy and phosphorus is a mined resource with limited quantities available worldwide. According to INSA Toulouse, which is studying several scenarios for wastewater resource recovery by means of collection at the source of the various resources that can be recovered, the one with the highest mitigation potential is nitrogen. Note that avoided emissions need to be accounted for separately from the GHG emissions of the services, in order to avoid double counting by the producer and the user of the recovered nutrients.

Beyond nutrients, biomaterials can be produced from wastewater, such the production of bioplastic by bacteria. These are lab or pilot scale innovations led by research and large wastewater treatment plants and are still to prove their feasibility for full scale implementation. Drinking water sludge may also be reused to recover materials[56] , such as lime pellets[57].

2.2   Low-carbon energy production

Water and wastewater services can provide a quantity of energy roughly equivalent to that which they consume during the treatment of water and wastewater[58]. We are therefore moving from the logic of elimination (waste) to that of recovery by considering wastewater and sludge as resources[59], and making the best use of gravity. Many solutions exist, each tailored to the size of services[60] . For this energy production to really be low-carbon, utilities need to ensure they control their nitrous oxide emissions and methane leaks[61].

Water and wastewater services may opt to sell or self-consume the energy they produce. Selling to a third party, such as the city heating district or gas grid, can be very relevant. However, it requires to account for the local regulatory requirements and to set sound contractual terms in order to secure the volume and unit prices of the transaction to ensure a return on the infrastructure investments.

---

56. IWA 2016, Water utility Pathways in a Circular Economy.

57. Dutch water Sector Article, 2016: Reuse of lime from drinking water softening is profitable in a wider range of products.

58, 59. Guide Amorce Services publics d'eau et d'assainissement et changement climatique   : leviers d'atténuation, 2019

60. Projets STEP du futur AERMC et laboratoire RESEED Ressources eaux et déchets INSA Lyon et IRSTEA

61. Guidelines on biogas leakage reduction by the European Biogas association

USING AVAILABLE LAND AND SURFACES TO PRODUCE SOLAR AND WIND ENERGY

Although not yet widespread, the production of solar energy on the buildings of water and sanitation utilities has an interesting potential. The energy produced can be partially self-consumed and partially sold. The installation of photovoltaic (PV) panels is also very relevant to provide the energy needed to pump groundwater in remote rural areas. This is the case, for example, of the project to provide water and electricity to 9 villages in the Plateaux region of Togo, supported by the NGO Electriciens sans frontières (see **Project #4-FWP**). Solar energy can also be coupled with other energy sources (wind, hydro, etc.) or installed in the form of floating photovoltaic panels on irrigation basins, reservoirs, artificial lakes, etc.

It is possible to use the water catchment areas and treatment buildings as a location for the installation of wind turbines. These facilities are easier to install in rural areas because of the acceptability constraints, as well as the fact that urban turbulence usually makes larger wind energy production in cities too inefficient. For the wind turbine to be viable, the installed power must be greater than 50 KW and the average wind speed higher than 3m/s at 50 m from the ground.

PRODUCING HEAT

Three approaches may be used to recover waste heat from wastewater:

- Recovery in the building where hot water is used, capturing wastewater before it enters the sewer network (integrated exchanger or at the foot of the building).
- Recovery within the sewage network on the existing pipes or new installations (see **Project #9-ASTEE**).
- Recovery at the wastewater treatment plant on treated water.
- Similarly, heat can be recovered from a groundwater supply, as in the example of the Albien groundwater storage, where Eau de Paris recovers heat to support local district heating (see **Project #2-ASTEE**).

In these installations, the critical success factor is identifying a local low temperature network or user (e.g. urban greenhouse) that could use the waste heat recovered. Indeed, as the system uses a heat pump to recover the heat, the temperature of the recovered hot loop is quite low, and may not be appropriate to all types of district heating, or on-site heating requirements.

PRODUCING HYDRO-ELECTRICITY

Hydraulic energy can be produced by installing micro-turbines to produce decentralized hydroelectricity, wherever the elevation drop and the flow rate are sufficient. For example the As Samra WWTP in Jordan has installed a hydroelectric turbine to support some of their electricity needs (see **Project #9-FWP**).

These installations have the most potential in large transport pipes, where the flows are high, even if the elevation drop is low, or in mountain areas, where the elevation drop is high even if the flows are not very big. The microturbine technology for drinking water is for example used in Papeete in French Polynesia by the Société Polynésienne des Eaux (see **Project #5-FWP**). This avoids the use of carbonated energies, frequent in insular areas. Bergen Water Utility in Norway also has two water turbines inside a mountain where water is transported from its source to the drinking water treatment plant. This production provides for 90% of the energy demand on the drinking water services.

Treatment facilities can also be designed as energy storage by including a double stage water retention facility. The water in the upper basin is turbined at times when the price of electricity is highest and collected in the lower basin to be pumped back to the upper basin when prices are cheap. The upper reservoir is therefore a particularly interesting energy reservoir for storing excess intermittent energy such as solar or wind power plants. Wastewater, grey water, rainwater or drinking water can be used for pumping and storage.

PRODUCING ENERGY FROM BIOSOLIDS

Biosolids, beyond being a source of nutrients or materials recovery mentioned above, is also a source of energy because of its easily degradable organic matter that can be transformed into methane or heat, through methanization or incineration respectively.

Below is an overview of different processes to recover energy from biosolids:

- The incineration of residual sludge coupled with heat recovery can supply industries or heating networks with high temperature heat. The heat can also be internally used for digestor heating, as in the example of Grenoble, France (See **project #1 ASTEE**).
- The anaerobic digestion of biosolids produces biogas through methanization. Biogas can be transformed into biomethane after purification, used as fuel to cover local heating requirements, or transformed into heat and electricity through a Combined Heat and Power engine. Anaerobic digestion is typically implemented for treatments plant of 30 000 PE or more. However, if biogas leakage is not monitored and controlled, methane emissions can negate the benefits of the biogas use in place of other fossil fuels[62].
- Hydrothermal processes[63],[64] (See **project #6 PFE**).
  > Pyrolysis is a thermal process that is more efficient than incineration to recover energy as bio-oil, biochar, syngas, and heat.
  > Gasification is a thermal process that enables to recover energy in the form of syngas and biochar.
  > Carbonization is a thermal process that produces biochar, which can be used as a fuel source.

Methanization is a biological process of degradation of organic matter in the absence of oxygen, allowing the production of biogas and digested sludge. This is what happens during the anaerobic digestion. Biogas is a gaseous product essentially composed of methane, carbon dioxide and water vapor. Once purified, it forms Biomethane (more than 97% methane) which can be injected into the natural gas distribution network or used as fuel for certain vehicles. The digested sludge is a solid wet product composed of residual organic matter not degraded by methanization and mineral matter. In the case of the Cometha Project developed in partnership between the wastewater and the solid waste utilities of the Greater Paris area, the digested sludge is transformed by means of thermal treatments (pyrolysis, gasification, carbonization…) to further recover energy

(see **Project #6 PFE**).

62. Danish Energy Agency report, 2021: Targeted efforts to reduce methane loss from Danish biogas plants ».
63. Water Research Foundation 2021, HYPOWERS: Hydrothermal Processing of Wastewater Solids Phase 1
64. Cométha innovation partnership SIAAP-Syctom,

These processes allow - under the action of heat and/or pressure, and under a controlled oxygen atmosphere - to transform part of the digested sludge into syngas, essentially composed of nitrogen, carbon dioxide, carbon monoxide, methane and hydrogen. This synthesis gas can be stored, used to produce heat or be converted into biogas through methanation[65]. The heat, biogas, biomethane, syngas, hydrogen or biochar recovered from biosolids treatment are considered renewable energies. They can be used for example by city heating or gas grids, or by public transport fleets. For example is the biogas from Bergen Utility in Norway, utilized as fuel in local buses, the energy production in 2021 was sufficient for to drive 100 times around the earth. However, the GHG emissions associated to their production have to be accounted for in the GHG assessment of the treatment plant, and only then can the avoided emissions associated to the use of the above listed renewable energies be highlighted in the assessment. Note that these avoided emissions need to be accounted for separately, to avoid double counting of the GHG emissions reduction associated to renewable energy use.

These solutions are most relevant when the direct recovery of organic matter and nutrient in the soil (composting, land application) are not possible in the vicinity of the facility or when the biosolids do not meet the quality requirement for land application. In cases where biosolids transport for land application is significant, the energy recovery solution may offer a better carbon footprint overall.

---

65. https://www.syctom-paris.fr/fileadmin/mediatheque/documentation/cometha/Syctom-Cometha_Dossier_2.pdf , page 5.

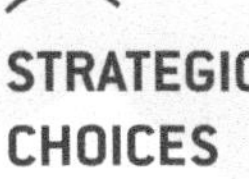

3. DECISIONS AND STRATEGY

Reducing GHG emissions doesn't only rely on technical measures, but also on the enabling environment created by the regulations and governance structures, as well as on a set of decisions and strategic positioning of the utility.

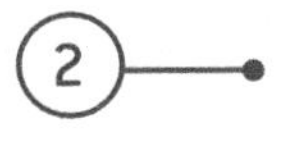

3.1 Awareness raising and education

Stakeholders at all levels - population, decision-makers, professionals- must be informed about climate issues and water management to enable them to take responsibility and ownership of the issues and facilitate decision-making. Scientific research and knowledge is essential to provide stakeholders with science-based information and data to support sound strategic planning and knowledge exchange.

At the level of water and sanitation utilities, this translates into diagnostic work and monitoring of specific indicators such as quality of water resources, withdrawals and consumption, or energy consumption and production.

The training of technical staff together with the awareness raising amongst the local population contributes to defining a roadmap to assess and monitor GHG emission, and set an action plan that involves end users to reduce emissions and adapt to the impacts of climate change. This is the

approach adopted by the Metropolis of Montpellier with the Aquamétro program (See **Project #2 FWP**). It consists of informational workshops on water savings, targeting different audiences: schoolchildren, parents, tenants and owners, municipal services, water and sanitation services.

3.2 Governance that supports changing practices

Mitigation and adaptation actions are guided by formal and informal governance around water-related issues. Indeed, governments are not the only ones with responsibility for water and sanitation services, especially in low-income countries where the share of non-state actors is growing. Increasing constraints and pressures on the water resource triggers competition for access to water. Its allocation is subject to negotiation. Improving governance is key to the sector's resilience to climate change. Water management in the context of climate change requires :

- greater public participation to discuss and manage climate risks
- Building capacity for mitigation and adaptation
- prioritizing risk reduction for the most vulnerable

However, adaptation and mitigation challenges also require specific infrastructure, information, legal and economic instruments to plan actions, and monitoring and coordination mechanisms, all of which require institutions.

Integrated water resource management makes it possible to involve the various stakeholders in climate issues at all levels of society, the economy and the environment (see **project #12-FWP**). Sectoral fragmentation can be an obstacle to the integration of these issues. Scientific information and data must be accessible locally in order to be integrated into local decision-making among stakeholders. Consultation is essential in the context of climate change to allow for a shared understanding of the challenges, the ownership of the proposed solutions and a common vision for the implementation of mitigation actions. Sharing experiences and lessons learned from projects is an accelerator to changing practices.

3.3   Responsible consumption and economic incentives

In many places, the price of water is heavily subsidized and therefore does not reflect the full cost of water and sanitation services, the difficulties of production in water-stressed areas, and does not take into account periods of scarcity. There is therefore very little incentive to conserve water and reduce consumption. To remedy this, various solutions have been devised:

- Progressive environmental and social water pricing provides an incentive to reduce water consumption while guaranteeing access to water for all. This measure has been implemented in France since 2010 in areas where there is a shortage of water and 50 local authorities have been authorized to experiment with social pricing (see **project #2-FWP**).

- Other incentives are based on the market economy but, if badly managed, can be ineffective or become harmful. This is the case with the creation of a water market, as in Australia and recently in the United States, which can lead to abuses such as speculation or the reinforcement of inequalities in access to water. The value of the water markets tool raises questions.

- Incentives to reduce GHG emissions of services can occur through a carbon tax or carbon credits. Large utilities in France may benefit of carbon credits when using biogas or waste heat in place of fuel to support their heating needs.

3.4.   Choosing low-carbon energy and supplies

The water and sanitation services can make a decision to further reduce their GHG emissions by reducing their scope 2 and 3 emissions, by purchasing green energy and adjusting the type of materials used as consumables for treatment. For example Eau de Paris is investigating the replacement of activated carbon by coco fiber to reduce its supply chain GHG emissions (see **project #4-ASTEE**).

These approaches need to be supported by decision support tools. For example the Carbon Calculator developed by local authorities in Seine Saint Denis enables to assess the impact of public works on GHG emissions, and adapt construction methods and materials selection to lower the impact of new construction or rehabilitations. Bergen Water Utility in Norway also has a self-developed climate footprint tool (including tier 2 and 3). Based on this tool the utility has changed e.g. filter materials and chemicals, on the background of their production emissions and transport emissions.

Working in collaboration with suppliers, utilities can make choices to purchase chemicals with lower-carbon footprints from both their fabrication and their transport.

GHG emission reductions *combined with* **adaptation** *and* **water resources preservation solutions**

1. PROTECTION OF WATER RESOURCES

Water and sanitation services are affected by the deterioration of water quality associated to reduced low flows and higher temperatures in rivers caused by climate change and other interrelated factors[66]. These might stimulate the growth of algae and bacteria in water bodies, while the solubility of oxygen might decrease.[67] These services are anticipated to have to spend more energy to treat degraded drinking water sources and meet more stringent discharge requirements for treated wastewater due to the increased risk of eutrophication of rivers and streams, for example.

Water services sometimes prefer to act upstream to preserve water resources. This is the case of **Eau de Paris**, which supplies drinking water to three million users and implements a policy of protecting water resources through partnerships with farmers to limit the use of pesticides and inputs into water reservoirs upstream. In 2020, this policy was reinforced through an innovative system of subsidies to farmers to compensate for the environmental services they provided. Another policy is the acquisition of land and its rental to farmers who are involved in environmental approaches through rural leases, "zero pesticide" policies, and ecological management of catchment areas.

Water resources may be preserved through nature based solutions such as restoring the connectivity of a waterway and its continuous flow, planting trees that produce shade and therefore prevent increasing water temperatures, or restoring ecosystem services such as natural filtration. All these solutions also contribute to stabilise or reduce GHG emissions, for example through carbon sequestration.

---

66. USGS Temperature and Water webplatform
67. *GIZ et al.* 2020: Stop Floating, Start Swimming, p. 25

2. ADAPTING SERVICES TO THE IMPACTS OF CLIMATE CHANGE

Beyond the protection of water resources at the source, water and sanitation services have to adapt their water supply and storage strategy, as well as their planning and operation.

2.1 Water supply *(drinking water services)*

Utilities are faced with the issue of water shortages and pollution which is becoming more and more frequent as a consequence land use change, population growth, increased urbanisation, growing demand etc, and exacerbated by the impacts of climate change. In order to guarantee supply and secure quality and quantity, utilities rely on the **interconnection of supply networks** and on **diversifying their resources** portfolio. Most important is to ensure a good knowledge of the resource capacity and to **plan to adapt abstraction to the evolving capacity of the resource**. During periods of quantity deficits preventive measures and the restriction of certain uses (watering, urban cleaning) are put in place by local authorities, but also by mobilizing citizens as mentioned above. These restrictive measures were put in place successfully in France, as well as in many other countries, such as in Australia during the Millennium drought, or in Cape Town more recently.

> Most important is to ensure a good knowledge of the resource capacity and to plan to adapt abstraction to the evolving capacity of the resource.

2.2 Storage

Water storage can provide an answer to multiple impacts of climate change: excess water during flooding or extreme precipitation can be captured and released in times of water scarcity. Climate change threatens the existence of natural storage systems, such as glaciers and natural wetlands, which are often a main component of a basin's water cycle, buffering variable water availability.[68]

> *Figure 5:* **The potential of different water storage systems for increased climate resilience**

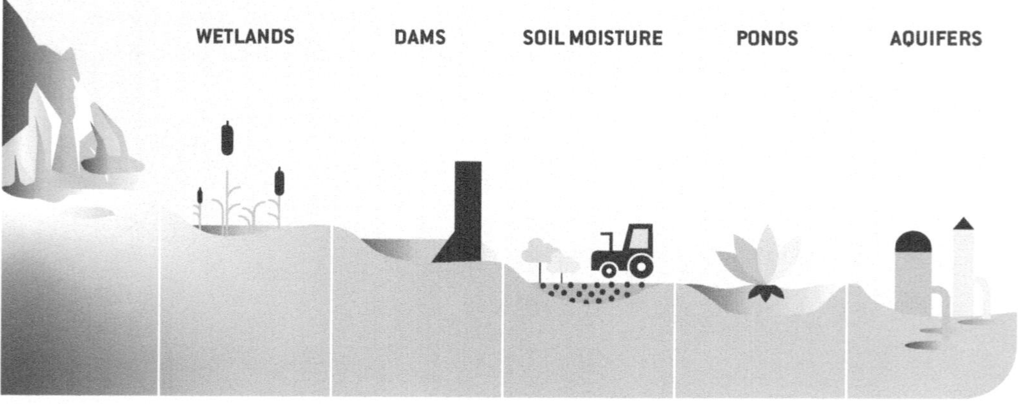

68. GIZ *et al.* 2020: Stop Floating, Start Swimming, p. 94

In a context of climate change, conventional grey infrastructure methods of water storage (reservoirs, dams) and water conveyance (water transfer through canals) are impacted by problems of silting and runoff, but also by concerns associated to their environmental impact resulting in a series of restrictions. Furthermore, the most feasible storage possibilities have often already been implemented (at least in developed countries). Faced with these limits, the UN[67], as well as IWMI[68] suggests that water supply could be assured through hybrid solutions associating conventional methods and nature-based solutions (NbS). To ensure the availability of water in a context of climate change, several measures appear necessary:

- **setting up emergency water storage** enabling the distribution of drinking water in case of extreme drought. For instance, the city of Paris has an underground storage system (Albien groundwater storage) the use of which today is restricted to situations of shortage. In the spirit of reducing GHGs emissions, such emergency storage should as much as possible be designed as gravity-fed or rely on energy-saving water towers that reduce pumping needs. In addition, urban synergies may be identified to use these emergency water storage as heat & cold reservoirs for local district heating systems, which is what has been implemented by Eau de Paris on the Albien groundwater storage (see **project #2-ASTEE**).

- **increasing the capacity of storage systems and their management**: As we have seen, there are limits to how much the capacity of conventional systems (dams, reservoirs) can be increased. Nonetheless, the use of NbS could contribute to an increased storage through designed wetlands, or increased water retention in the soil.[69]

- **Managed aquifer recharge** (groundwater) is intended to increase the available groundwater volume by enhancing the natural water infiltration into the aquifer. This can be rainwater or treated wastewater. Aquifer recharge has many benefits, including optimization of water storage, replenishing of emptying aquifers, improving degrading groundwater quality, limiting the risks of saltwater infiltration into coastal aquifers or sinkholes. In addition, aquifer recharge may also be combined with a flood control strategy through retention ponds. NbS such as protecting or revegetating soils that foster water infiltration is a very efficient low cost solution to improve the replenishing of aquifers, Implementing these NbS in large scale requires partnerships which take time to build, but may deliver benefits beyond the sole replenishing of aquifers.

Windhoek, Namibia's capital city, is located in one of the driest areas of southern Africa, where the combined lack of rain, strong evaporation, and drinking water supply abstraction prevent the groundwater table from recharging. Given this permanent situation of water stress, as of 1968 the water system operator implemented wastewater recycling to recharge the groundwater that supplies drinking water (see **project #10-FWP**).

67. World Water Development Report 2020- L'Eau et le changement climatique, 2020
68. IWMI/GWP 2021: Storing water: A new integrated approach for resilient development.
69. Nature-based Solutions for Water 2018: The United Nations World Water Development Report 2018

2.3   Adapting operation of water and sanitation services

The complementarity of adaptation and mitigation in the face of climate change appears through the way water and sanitation services adapt their planning and operation to best meet their evolving needs. The following approaches illustrate this complementarity:

- **Adapting the size of infrastructure:** Setting up modular infrastructure that can be adapted to varying contexts, and enable to use less energy when possible. It's about being more adaptive to future uncertainties and avoid investing in oversized infrastructure that may never run at their optimal efficiency level. Leveraging NbS in hybrid systems may enable this enhanced adaptability.

- **Adapting the treatment with a fit-for-purpose approach and deliver only the quality of water required for the specific use.** In water scarce areas, treated or partially treated wastewater can be used to water crops[70], green spaces[71], clean public areas, or as industrial water instead of being discharged to the environment, in particular when the energy required to meet the requirements for these usages is lower than the energy required to discharge and identify other water supplies. Note that in the context of increasing river temperatures, the energy consumption associated to nitrogen removal from wastewater may increase to meet the environmental discharge requirements.

- **Preventing flood risks and managing rainwater**

  Sound rainwater management enables to reduce 1/ downstream runoff in streams and rivers and 2/ the discharge in collective sanitation infrastructure such as sewers, collectors, storage or treatment plants. It results in reduced flooding and mudslides risks, as well as in a reduced number of overflow events where large amounts of untreated water are discharged into rivers. Reducing the dilution of wastewater is also an opportunity to reduce the level of uncertainty that results in oversizing when designing a treatment plant, as well as a solution to reduce the energy consumption required to remove a specific pollutant load.

Treating diluted wastewater increases the volume treated and makes the removal of the same pollutant quantity more energy intensive. Beyond the sound management of rainwater, utilities and their cities reduce their vulnerability by putting in place an emergency alert system and preventive information, but also by identifying solutions to limit risks on their most vulnerable assets.

> Beyond the sound management of rainwater, utilities and their cities reduce their vulnerability by putting in place an emergency alert system and preventive information

---

70. <u>NbS for Wastewater treatment publication</u> on p. 27 and 34.

71. In France, only "watering green spaces" is authorized using treated wastewater, with or without additional treatment, depending on the required level of quality (A, B, C or D, cf. Ministerial decree dated 06/08/10).

An array of solutions can be implemented to reduce these risks. One is to switch from a combined-sewer network to a **separated network**, collecting wastewater separately from rainwater. This is an expensive solution but allows for different treatments tailored to different water characteristics. When a separated network is not feasible, combined sewer overflows may be treated through a wetland system, which is cost efficient and requires less energy than mechanical treatment.

Another solution is **integrated management of rainwater** and de-waterproofing. The problem of flooding from rainwater is strongly correlated to the reduced permeability of soils transformed through urbanisation and preventing rainwater from soaking in where it falls. In France, approximately 200-250 km² are turned into non-permeable surfaces every 25 to 30 years. And nearly 80% of the pollution of a drop of water is caused during runoff. It is therefore particularly relevant to encourage local infiltration of rain water.

The approach of Roannaise des Eaux (Loire) is particularly inspiring for the integrated management of rainwater (see **Project #12-FWP**). The solution of a separate network being too expensive for the local authority, it was decided to set up specific measures to manage rainwater following a set of priorities:

- **limiting non-permeable surfaces** during construction and, when possible, building upwards.
- turning unbuilt areas into green spaces to increase infiltration rates
- building access roads and parking lots with permeable surfaces
- encouraging rainwater retention on rooftops: green roofs, terraces on rooftops, etc.
- using nature-based solutions such as: valley gutters, dry ponds, draining trenches, infiltration wells, rain gardens, etc.
- **Disconnect and make urban surfaces permeable**: a pro-active program aiming to progressively de-waterproof and disconnect rainwater in existing urban spaces, taking advantage of the synergy of on-going projects (roads, buildings…).

SPOTLIGHT on Ecosystem services and nature-based solutions solutions (NbS)

Nature-based solutions (NbS) can be relevant for water and sanitation services in the context of climate change mitigation and adaptation. The International Union for the Conservation of Nature (IUCN) defines them as "actions to protect, sustainably manage, and restore natural and modified ecosystems that address societal challenges effectively and adaptively, simultaneously providing human well-being and biodiversity benefits". NbS can be implemented in all types of environment: rural, urban and natural. They provide ecosystem services that can be a real advantage for water and sanitation services. The solutions that can be mobilized to benefit water and sanitation services belong to various categories of NbS:

> **Ecological restoration:** supports ecosystems function better through hydro-morphologic measures, that improve both the natural purification capacities of water and, infiltration, which in turn prevents flooding.

> **Ecological engineering** may be used for alternative wastewater management, or after treatment by sanitation services. Examples include filtering gardens (reed beds or other plants depending on the local climate), natural lagooning, vegetated discharge areas, etc.

Filtering gardens as a purification solution were experimented for example in Morocco's Tiznit province. Pollutants and heavy metals are concentrated in the filtering plants and the water is purified. The plants are renewed and after they are incinerated, the ashes are reused (eg. in cement plants). Note that biomass incineration produces mostly biogenic CO_2 emissions and very small amounts of N_2O. Direct GHG emissions (CH_4 and N_2O) may occur during the treatment within the wetland but are typically compensated by the carbon sequestration in the biomass. Natural lagooning or the use of artificial wetlands are also a natural self-purification technique: a small flow of water is gradually purified as it passes through a series of buffer basins.

These intermediate zones help protect the receiving environment from the direct discharge of wastewater, either as a substitute to a wastewater treatment plant, or as tertiary treatment before discharge into the natural environment in particularly sensitive areas.

> **Green infrastructures** (networks of green spaces, ecological corridors) which are part of integrated rainwater management may become urban biodiversity reserves through rain gardens, swales, etc.

> **Natural water retention measures** are actions that contribute to slowing down water run-off by restoring ecosystems and changing agricultural and forestry practices. These actions positively impact water services by reducing flooding risks and the quantitative deficit of water bodies. They take on various forms in rural environments (hedgerows, weirs, meadows), in urban settings (vegetal roofs, retention ponds) or in natural environments (restored wetland, renatured watercourses, etc.).

---

72. Badiou *et al.* 2011: Greenhouse gas emissions and carbon sequestration potential in restored wetlands of the Canadian prairie pothole region.

5 French expertise: examples of solutions

This chapter presents 13 practical cases carried out in France or internationally by French actors and partners of the French Water Partnership (FWP).

These case studies illustrate the three main categories of actions for mitigation and adaptation: lean and efficient approaches, circular economy, and strategic choices, which have been presented in the previous sections.

In parallel to the development of this report by the FWP, in partnership with the International Water Association (IWA), the french Scientific and Technical Association for Water and the Environment (ASTEE) produced a publication on the challenges of reducing greenhouse gas emissions in water and sanitation utilities in France. The 11 case studies compiled by ASTEE are presented in the below table. The full description of these cases is available on ASTEE's website.

N° FWP

N°	
2	*Montpellier, France* - **Aquamétro, program to raise awareness about water savings among various groups** Locale water and energy agency Montpellier Metropolitan Area
3	*Chiang Mai, Thailand* - **Assessment of GHG emissions and action to prevent leakages, optimisation of pumps** Waste water management authority, IWA GIZ, financed by German Ministry of Environment
4	*Plateaux Region, Togo* – **Solar pumping, ecological sanitation, organization of local governance committees** Electriciens sans frontières (Electricians without borders)
11	*Ouijjane, Tiznit Province, Morocco* — **Sanitation with reed bed filters and recycling for agriculture,** Association Migrations & Development
12	*Roannais, France* — **Integrated management of rainwater** Roannaise water syndicate
13	*Paris, France* - **Protecting the water resource for less treatment** Eau de Paris

N° Astee

N°	
3	*Grand Chalon, France* – **Decarbonation,** Suez and Climat Mundi
5	*Paris, France* - **Energy savings,** Eau de Paris
6	*Grenoble, France* - **Hydraulic optimisation,** Grenoble Alpes Métropole
7	*Grenoble, France* - **Pretreatment optimisation,** Grenoble Alpes Métropole
8	*Bas-Rhin, France* - **Non-collective sanitation,** Aquatiris and Alternative Carbone
10	*Le Havre, France* - **Inciniration optimisation,** Le Havre Seine Métropole and Veolia
11	*Ile-de-France, France* - **Optimized pumps,** SAUR

CIRCULAR ECONOMY

N° FWP

N°	
1	*Cusco, Peru* - **Reducing GHG emissions from urban wastewater management** SEDACUSCO IWA GIZ, supported by German Ministry of Environment
4	*Plateaux Region, Togo* – **Solar pumping, ecological sanitation, organization of local governance committees** Electricians without borders
5	*Papeete, French Polynesia* – **Electricity production from micro-turbines in drinking water** Polynesian water company
6	*Paris, France* - **Cométha: pilot project for co-methanation of a non-reusable portion of organic waste sourced from household waste and sewage sludge** SIAAP SYCTOM
7	*Gaza, Palestine* – **Reusing gray water at the source in households for irrigation,** Secours Islamique France (NGO)
8	*Windhoek, Namibia* – **Recycling wastewater into drinking water, replenishing water tables** Veolia
9	*Amman, Jordan* - **As Samra Own-energy production, reuse of wastewater for agriculture** SUEZ
10	*Durban, eThekwini Municipality, South Africa* - **Water resilient city, recycling wastewater, preserving water resources** Veolia eThekwini Municipality
11	*Ouijjane, Tiznit Province, Morocco* – **Sanitation with reed bed filters and recycling for agriculture** Association Migrations & Development

N° Astee

N°	
1	*Grenoble, France* - **Biogas production,** Grenoble Alpes Métropole
2	*Ile-de-France, France* - **Geothermal heat,** Eau de Paris
9	*Issy-Les-Moulineaux, France* - **Heat from wastewater,** Veolia and Issy Energies Vertes

STRATEGIC CHOICES

N° FWP

N°	
1	*Cusco, Peru* - **Reducing GHG emissions from urban wastewater management** SEDACUSCO IWA GIZ, supported by German Ministry of Environment
2	*Montpellier, France* - **Aquamétro, program to raise awareness about water savings among various groups** Locale water and energy agency Montpellier Metropolitan Area
6	*Ile de France, France* - **Cométha: pilot project for co-methanation of a non-reusable portion of organic waste sourced from household waste and sewage sludge** SIAAP SYCTOM
12	*Roannais, France* - **Integrated management of rainwater** Roannaise water

N° Astee

N°	
4	*Paris, France* - **Low carbon reagent,** Eau de Paris

(| Mitigation (| Adaptation

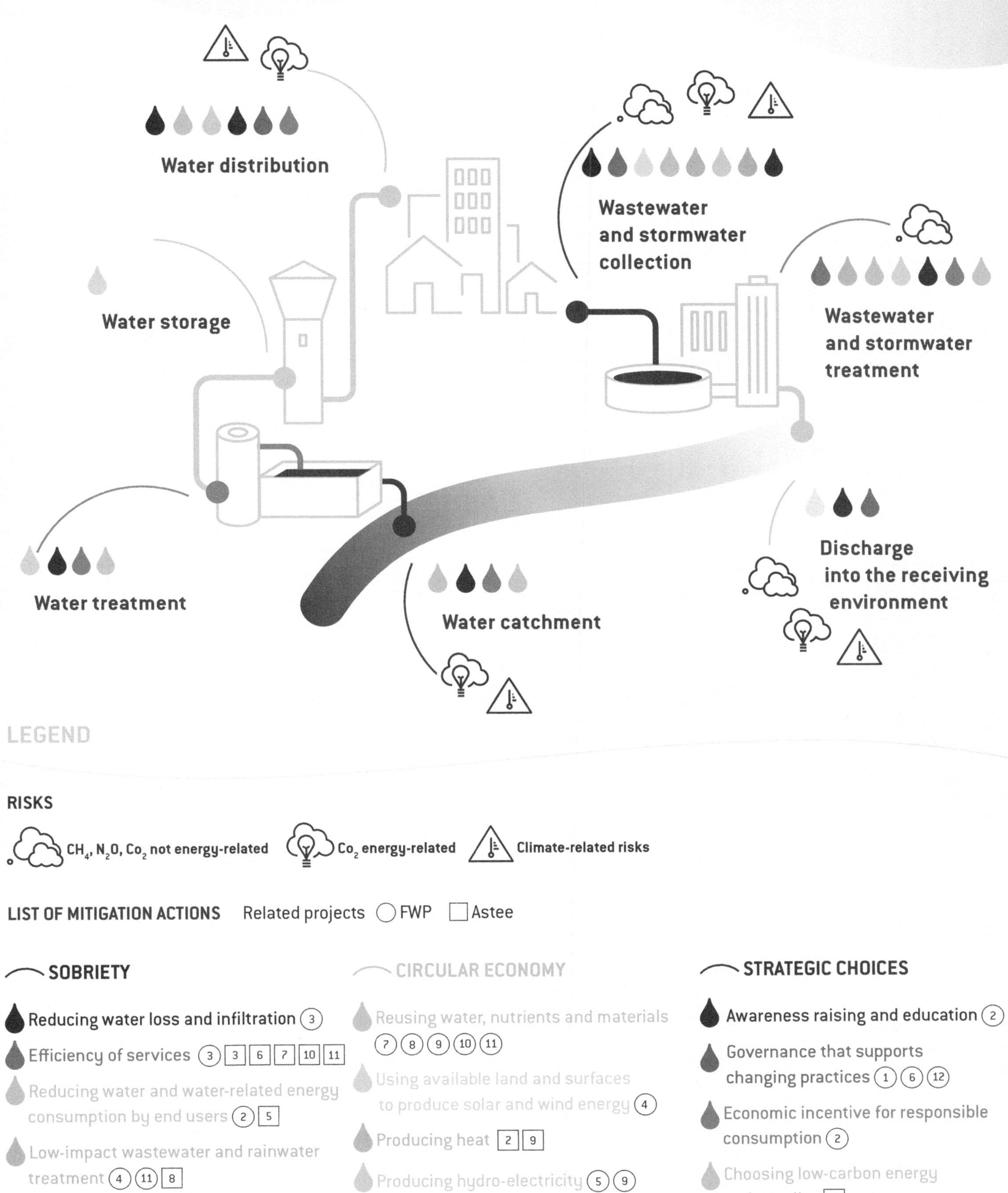

Reducing GHGs
from water and sanitation services

Water distribution

Water storage

Water treatment

Wastewater and stormwater collection

Wastewater and stormwater treatment

Water catchment

Discharge into the receiving environment

LEGEND

RISKS

CH4, N2O, Co2 not energy-related

Co2 energy-related

Climate-related risks

LIST OF MITIGATION ACTIONS   Related projects   FWP   Astee

SOBRIETY
Reducing water loss and infiltration 3
Efficiency of services 3 3 6 7 10 11
Reducing water and water-related energy consumption by end users 2 5
Low-impact wastewater and rainwater treatment 4 11 8
Alternative rainwater management 12
Restoring and preserving the quality and quantity of water resources 13

CIRCULAR ECONOMY
Reusing water, nutrients and materials 7 8 9 10 11
Using available land and surfaces to produce solar and wind energy 4
Producing heat 2 9
Producing hydro-electricity 5 9
Producing energy from biosolids 1 6 1

STRATEGIC CHOICES
Awareness raising and education 2
Governance that supports changing practices 1 6 12
Economic incentive for responsible consumption 2
Choosing low-carbon energy and supplies 4

1

Producing clean energy and reducing GHG emissions through improved wastewater management in Cusco

SECTOR:
Water and Sanitation

PROJECT PROMOTER:
Peruvian Ministry of Housing, Construction and Sanitation (MVCS) and German Federal Ministry for the Environment, Nature Conservation, Nuclear Safety and Consumer Protection (BMUV), implemented through the WaCCliM project by GIZ and IWA

CONTEXT

LOCATION: Cusco, Peru

TYPE OF TERRITORY: URBAN

CONTEXT PRIOR TO ACTION: In Cusco, sewage treatment by the municipal water and wastewater utility SEDACUSCO has generated about 110,000 tons of sludge each year. Without proper treatment, the sludge attracted insects, led to a serious odor problem and contributed to environmental degradation, while the resulting methane emissions as well as the emissions from energy use accelerated climate change.

BENEFICIARIES:

around 350,000 inhabitants

PROJECT DESCRIPTION

With support from MVCS and BMUV, the SEDACUSCO utility installed and started operating an anaerobic digester for treating sludge and producing biogas, which is however flared without valorization. As a next step, SEDACUSCO is planning to turn the produced biogas into thermal and electric energy, which will be used for operating the wastewater treatment plant, saving energy costs and CO_2 emissions and making the treatment plant's energy supply self-sufficient.

COST OF THE PROJECT: n/a

IMPACT ON EMISSIONS: The amount of GHG emissions resulting from untreated sludge was reduced by more than 7,400 tons of CO_2 equivalent per year, which corresponds to more than 5,500 passenger flights from Lima to Frankfurt and back. The operation of the biogas-powered clean energy production system will save another estimated 544 tons of avoided carbon per year as well as EUR 260,000 in annual electricity costs.

2

Aquamétro, common strategy for saving water and energy

SECTOR:
Action targeted at consumers
(water distribution stage)

PROJECT PROMOTER:
Local energy and climate agency Montpellier Metropolitan Area

CONTEXT

LOCATION: Montpellier Metropolitan Area, France

TYPE OF TERRITORY: URBAN / SUBURBAN

CONTEXT PRIOR TO ACTION:

- **88,622 m³ distributed daily,**
- **183L/pers/day,**
- **network efficiency is 83.2%**

BENEFICIARIES:

28 towns out of the 31 towns of the Montpellier Metropolitan area, Population Montpellier

METROPOLITAN AREA:

481,276 inhabitants.

PROJECT DESCRIPTION

PROJECT DATES: 2016 - today

- **GHG emission sources concerned:** emissions related to energy scope 2 through water savings

- **Description of the actions to reduce GHG emissions:**
> Action 1: Water consumption of municipal estate
> Action 2: Data base of water consumption
> Action 3: "éco'minots" challenge
> Action 4: Supporting water savings among the general public
> Action 5: Supporting water savings in collective housing ("Copr'eau")
> Action 6 (forthcoming): "Water efficient town" certification
> Action 7: Parents challenge

COST OF THE PROJECT: 413,000€ between 2016 and 2020 including 2 FTE

MAGNITUDE OF WATER SAVINGS:
Action 1 = 17% water savings obtained in particular through early detection of leakages
Action 3 = éco'minots challenge: 20% water savings on average

Reducing the greenhouse gas emissions of the sanitation services in Chiang Mai, Thailand

3

SECTOR:
Wastewater treatment

PROJECT PROMOTER:
WaCCliM, implemented by GIZ and IWA on behalf of German Federal Ministry for the Environment, Nature Conservation, Nuclear Safety and Consumer Protection (BMUV), Thailand's Wastewater Management Authority (WMA) 1, Thailand's Ministry of Natural Resources and Environment (MNRE)

CONTEXT

LOCATION: Chiang Mai, Thailand, Ping River northern area

STATE OF THE WATER IN THE RIVER BASIN: polluted river

TYPE OF TERRITORY: URBAN

CONTEXT PRIOR TO ACTION: In the city of Chiang Mai, a baseline study has shown that water pollution is mainly caused by leakages in the city's wastewater collection system, flowing directly into the canal. These flows are also the main source of the utility's GHG emissions (methane and nitrous oxide, which have extremely high warming potential). The emissions resulting from the direct discharge of untreated water were estimated at 579,900kg of CO_2 equivalent in the city. In addition to the GHG emissions, the water in the Ping River suffers from environmental pollution through the wastewater effluent.

BENEFICIARIES:

Among a population of 137,000, 50% was connected to sewers and about 30% of collected wastewater was treated in the water treatment plant.

PROJECT DESCRIPTION

PROJECT DATES: 2014 - 2018

- **GHG emission sources concerned:** energy, as water savings will decrease volumes

- **Description of the actions to reduce GHG emissions:**
> The WaCCliM project's ECAM tool was used to identify direct and indirect sources of CO_2, CH_4 et N_2O at all treatment stages. Optimizing pump energy efficiency and fixing leaks should result in at least a 12% reduction in emissions. In addition, the cooperation between WaCCliM and the WMA in Thailand has raised the local awareness for the challenges in the wastewater sector and the need for improvements in the urban water management.

COST OF THE PROJECT: n/a

Villages, health and sustainable development

4

SECTOR:
Capture, treatment and decentralized distribution, ecological sanitation

PROJECT PROMOTER:
Electriciens sans frontières (Electricians without borders)

CONTEXT

LOCATION: Plateaux region, near the city of Notsé, Togo

TYPE OF TERRITORY: RURAL

CONTEXT PRIOR TO ACTION: No access to water, sanitation or electricity in 9 villages

BENEFICIARIES:
85,000 direct or indirect beneficiaries, 9 rural villages, 3 secondary schools (1 lycée, 2 collèges), 3 primary schools, 6 medical dispensaries

PROJECT DESCRIPTION

- **GHG emission sources concerned:** emissions related to energy scope 2 and to treatment scope 1

- **Description of the actions to reduce GHG emissions:**
> The mitigation actions taken as part of the project are the installation of solar pumps to extract water and of double-pit latrines with sanitation with filtering plants, and a dry well to enable the reuse of sludge for agriculture. Actions were taken to collect waste and organize its management in order to preserve watercourses and soils.

COST OF THE PROJECT: 720 000€+ 300 000€ worth of volunteer work

5 Vaimamara, electricity produced from turbines powered by drinking water

SECTOR:
Water service

PROJECT PROMOTER:
Société polynésienne des eaux (Polynesian water company, a SUEZ subsidiary)

CONTEXT

LOCATION: Titioro valley, Papeete,Tahiti. French Polynesia, France

TYPE OF TERRITORY: CITY

CONTEXT PRIOR TO ACTION:
70% of electricity in Polynesia comes from fossil fuels. It was the case for Papeete's water service. Using fossil fuels also poses risks of pollution of the terrestrial and marine environment, especially of coral reefs, and as the price of fossil fuels increases, so does the price of water for the inhabitants.

BENEFICIARIES:
30,000 inhabitants of Papeete

PROJECT DESCRIPTION

- **GHG emission sources concerned:** reduction in emissions related to energy scope 2 because of substitution with hydroelectricity

- **Description of the actions to reduce GHG emissions:**
> The goal is to cover 80% of the water service's electricity requirements with renewable energy (supply 130 MWh/year out of 160 MWh/year for service HQ). Hydroelectricity produced with micro-turbines replaces the use of fossil fuels the electricity required by the water service to produce and distribute drinking water. This decreases the cost of drinking water (it was around 0.20€/kWh with fossil fuels). The goal is to stabilize the price when fossil fuels prices are likely to increase. Use of two 15kW and 30kW pump-as-turbine micro-turbines.

COST OF THE PROJECT: 144 000€ overseas tax reduction 17 000€

6 Cométha

SECTOR:
Recycling sewage sludge

PROJECT PROMOTER:
SIAAP SYCTOM

CONTEXT

LOCATION: "Seine Aval" SIAAP plant, Ile-de-France, France

TYPE OF TERRITORY: URBAN

CONTEXT PRIOR TO ACTION:
The non-reusable portion of organic waste is usually incinerated with household waste and most of the sludge is spread in fields.

BENEFICIARIES:
the plant has a capacity of 750,000 population equivalent

PROJECT DESCRIPTION

- **GHG emission sources concerned:** emissions avoided through sludge recycling (Scope 3)

- **Description of the actions to reduce GHG emissions:**
> The action is to put together a non-reusable portion of the organic waste from household waste, wastewater treatment and horse manure from Maison Lafitte to turn it into biogas which after purification may be recycled under other forms (biomethane and thermal energy). This biogas is injected into the urban grid.

PHASE 1 2018 2019: research and testing Phase 2 2020: design, construction and operation of 2 pilot units for treatment by co-methanation.

PROJECT DATES: 2017-ongoing

COST OF THE PROJECT: studies + R&D + pilot units = 23 M€. Project funded in equal parts by Siaap and Syctom. Funding from the Seine Normandie Water agency is 17,600€.

Grey Water Project , Gaza, Palestine

7

SECTOR:
Wastewater treatment
[treatment of gray water at the source, reuse]

PROJECT PROMOTER:
Secours Islamique France

CONTEXT

LOCATION: Governorates of Gaza, North Gaza, Dair al Balah, Khan Younis, Rafah, Palestine

TYPE OF TERRITORY: RURAL, ISOLATED

CONTEXT PRIOR TO ACTION: Tap water is not drinkable (damaged infrastructure, irregular electricity supply). 90% of households in Gaza buy water made from desalinated seawater (10 to 30 times more expensive than tap water and of low quality). Families have meagre means, they live off vegetable gardens that guarantee their food security and provide an income. The project is located in areas not served by the public sanitation network.

BENEFICIARIES:

320 households in total, meaning 1,824 persons equipped (100 households in 2021)

The water savings generated by the gray water treatment unit are estimated to be 64% (460 liters recycled out of the 720 consumed daily per household).

PROJECT DESCRIPTION

- **GHG emission sources concerned:** reducing emissions related to energy (scope 1) because of water savings and emissions related to treatment (scope 1) because lower volumes are treated in plants.

- **Description of the actions to reduce GHG emissions:**
> Goal 1: Increase the recycling of gray water and its reuse by households. Goal 2: Promote knowledge and added value from gray water treatment among communities, WASH organizations and local authorities and institutions. The project reduces discharges into the environment, the volumes of wastewater going through water treatment plants in Gaza due to treatment at the source, and floods due to water treatment plant overload.

COST OF THE PROJECT: 644,013 € including 203,514€ for the third phase in 2021

MAGNITUDE OF THE ENERGY SAVING estimated to be 153kWh
MAGNITUDE OF THE SAVINGS avoided: 68.8kg. CO$_2$ eq

Toilet to tap

8

SECTOR:
Wastewater treatment

PROJECT PROMOTER:
Veolia

CONTEXT

LOCATION: Windhoek, Namibia

TYPE OF TERRITORY: URBAN

CONTEXT PRIOR TO ACTION: The city of Windhoek lies in an extremely dry area. Underground water supplies 40% of the country's needs, water is also transferred from 3 dams. The demand for water is increasing. Conventional sources were no longer able to meet water requirements as early as the 1960s, so the city tried to reduce the demand and increase available resources (storage in dams, reuse of water, and optimized management of aquifers).

BENEFICIARIES:

400,000 inhabitants supplied with drinking water (greater Windhoek area)

PROJECT DESCRIPTION

- **GHG emission sources concerned:** reducing emissions related to energy (scope 2) and treatment (scope 1)

- **Description of the actions to reduce GHG emissions:**
> In order to reduce the demand for water, the city aimed to reduce losses in the distribution system and to adapt treatment to requirements in terms of water quality. It carried out communication campaigns about water savings, set up an incentivizing, progressive pricing system and restricts water usage during periods of water stress. The city has a dual network system with a network for non-drinking water. Any surplus from wet years is injected into the aquifer below the city as a reserve for dry years (managed aquifer recharge, MAR). This solution covers 35% of the city's needs and helps fight chronic shortages.

COST OF THE PROJECT: In 2001, investments (CAPEX et OPEX) for the two services were estimated to be between 3-27/m^3 N$ and 19/20m^3 N$ in 2018 for Namwater and between 5-71m^3 N$ and 13/00m^3 N$ in 2018 for Goreangab. The cost of the plant was estimated at 110 million in 2001, now approximately 400 million.

9 As Samra

SECTOR:
Wastewater treatment

PROJECT PROMOTER:
Suez

CONTEXT

LOCATION: Greater Amman, Jordan

TYPE OF TERRITORY: URBAN

CONTEXT PRIOR TO ACTION: The area is experiencing water shortages and high population growth. Amman lies in the middle of the desert. The plant meets the needs of the population, of agricultural and industrial activity.

BENEFICIARIES:
2.2 million inhabitants; 3.5 million projected in 2025.

PROJECT DESCRIPTION

- **GHG emission sources concerned:** reducing emissions related to energy (scope 2) and treatment (scope 1)
- **Description of the actions to reduce GHG emissions:**
> The plant underwent extension in 2012 and its capacity went from 267,000 to 364,000 m^3/day. The plant recycles wastewater and produces water of sufficient quality to be reused for irrigation. The plant produces renewable energy (hydroelectric turbines) on-site and the sludge is recycled into biogas. The plant produces 80% of its energy needs. Residual sludge is turned into pellets and reused as fuel or agricultural fertilizer. The plant covers approximately 10% of Jordan's agricultural needs. When treatment processes were optimized, the Zarqa River's water quality was restored.

COST OF THE PROJECT: 169 million dollars (phase 1) + 267 million USD (phase 2).

MAGNITUDE OF THE GHG EMISSION REDUCTION: Approximately 300,000 ton of CO$_2$ are saved every year as the plant produces renewable energy.

10 Water Resilient city

SECTOR:
Sanitation and drinking water

PROJECT PROMOTER:
Veolia, eThekwini Municipality

CONTEXT

LOCATION: Durban, South Africa, uMngeni river basin, eThekwinin municipality, KwaZulu-Natal administrative unit.

TYPE OF TERRITORY: URBAN

CONTEXT PRIOR TO ACTION: Durban is in a situation of water stress, dam reservoirs regularly are filled 20% below average. One quarter of inhabitants live in informal housing with uncertain water access. This poor population is particularly vulnerable to shocks, climate stress and natural disasters. "Business as usual" solutions such as building new dams or treatment plants, or transferring water from other basins have reached their limits when addressing these issues.

BENEFICIARIES:
3.158 000 inhabitants of eThekwini Municipality

PROJECT DESCRIPTION

- **GHG emission sources concerned:** reducing emissions related to energy (scope 2) and treatment (scope 1)
- **Description of the actions to reduce GHG emissions:**
> Veolia recycles 98% of the wastewater from the SWTW plant, which is then used by local industries for their own production processes. This reduces freshwater abstraction. Durban has also set up an inter-city partnership mobilizing several stakeholders to implement biodiversity preservation and adaptation measures.
The goal of these measures is to preserve water resources and limit the effect of natural disasters by restoring ecological infrastructures and protecting ecosystem services. The water and sanitation unit initially led this initiative for the integrated management of biodiversity, climate and poverty issues at the level of the bio-region's social and ecological system. The aim was to improve the system's resilience.

MAGNITUDE OF THE WATER AND ENERGY SAVINGS: 47, 000 m^3 of freshwater saved. By recycling wastewater, local industries save more than 5 million euros annually.

Ecological sanitation in the province of Tiznit, Morocco

11

SECTOR:
Wastewater treatment

PROJECT PROMOTER:
Association Migrations & Development

CONTEXT

LOCATION: Villages of Assaka and Akal Melloulne in the Ouijjane commune, Province of Tiznit, Morocco

TYPE OF TERRITORY: **RURAL, ISOLATED**

CONTEXT PRIOR TO ACTION: **All the villages were connected to drinking water in 2014, but in the absence of sanitation systems, this created problems such as the increase of wastewater discharge into the natural environment close to villages. This in turn generates pollution risks for water tables, watercourses, soils, and increases waterborne diseases.**

BENEFICIARIES:
944 beneficiaries (190 households)

PROJECT DESCRIPTION

PROJECT DATES: **2017-2020**

- **GHG emission sources concerned:** reducing emissions related to treatment (scope 1) through an ecological sanitation system (energy savings, decreased use of reagents...).

- **Description of the actions to reduce GHG emissions:**
> The sanitation system that was chosen for this project, reed bed filters, is not widely used in Morocco yet. This solution decreases the health and environmental pressures exerted by human populations and the natural environment.
It also enables a profitable reuse of wastewater to develop local agriculture. Planted filters are a solution suited to rural towns of less than 2,000 inhabitants. This solution does not require a large budget and the reeds may be reused when cut.

COST OF THE PROJECT: 590,937 €

Developing a voluntarist policy for integrated rainwater management

12

SECTOR:
Rainwater management, sanitation

PROJECT PROMOTER:
Roannaise de l'eau

CONTEXT

LOCATION: Le Roannais, Loire, France

TYPE OF TERRITORY: **URBAN/SUBURBAN/ RURAL**

CONTEXT PRIOR TO ACTION: **This policy came as an answer to problems linked to flooding risks such as the overflow of stormwater basins. In Roanne, 20% of effluent water is not treated when it rains. This is a major source of water pollution. The cost of treating them (pollution basins) was estimated at 47 million euros which was too expensive for the local authority governing the greater Roanne area, considering other solutions exist.**

BENEFICIARIES:
102,574 inhabitants on the territory of la Roannaise des eaux, 40 towns

PROJECT DESCRIPTION

- **GHG emission sources concerned:** reduction of emissions related to treatment (scope 1) through alternative rainwater treatment with NbSs

- **Description of the actions to reduce GHG emissions:**
> Rather than treating rainwater, the idea is to limit run-off as much as possible by decreasing impermeable surfaces and thus allowing water to infiltrate where it falls. The goal of the syndicate is to disconnect 32 hectares of impermeable surfaces over 10 years. The syndicate also focuses on raising the general public's awareness of rainwater management issues and infrastructure (information panels). Finally, "Responsible water" guidelines have been drawn up to help local authorities design water management strategies.
15 municipalities in the syndicate have committed to this.

COST OF THE PROJECT: by giving priority to disconnection and de-impermeabilisation operations, the cost of the action plan aiming to align Roanne's sanitation system with standards was reduced from 47 M€ to 32 M€.

13   Protecting the water resource for less treatment

SECTOR:
Production

PROJECT PROMOTER:
Eau de Paris

CONTEXT

LOCATION: Regions of Bourgogne-Franche-Comté, Grand-Est, Normandie, Centre-Val de Loire, Île-de-France, France

STATUS OF THE BODIES OF WATER CATCHMENT: Local qualitative problems (certain pollutants: nitrates and pesticides).

TYPE OF TERRITORY: SUPPLYING A HYPER-URBANIZED CITY (PARIS) but the projects of water catchment are located in rural areas.

CONTEXT PRIOR TO ACTION: Since 1990 actions in favour of resource protection are led by Water of Paris, and especially since the creation of the management department.

BENEFICIARIES:

3 million daily users

PROJECT DESCRIPTION

DATES OF THE PROJECT: **2021-2026**

- **GHG emission sources concerned:** reduction of GHG related to industrial potabilisation and sequestration of carbon from natural areas.

- **Description of the actions to reduce GHG emissions:**
  > Sequestration of carbon by maintaining or converting into grassland, or the conversion of areas of high ecological in-terest (hedgerows, buffer wetlands) ;
  > Reduction of emissions related to industrial activities, a better quality from the water spring allows a reduction of treatment by potabilization and the emissions associated.

COST OF THE PROJECT: The plan of agriculture assistance represents 47 million euros of annual support (including 37 -million euros financed by the Water Agency Seine Normandie, and 10 million euros by Water of Paris).

EMPIRICAL FEEDBACK: **Around 150 farmers committed with Water of Paris (biological agriculture, reduction of farm inputs, increasing grasslands, diversifying cultures, planting of hedgerows).**

Astee PROJECTS

Sheet n°	Title	Project carrier	Type of action			Greenhouse Reduction	Type of project	Investment cost	Reproductibility
			Technical	Organizational	Behavioral				
1	**Biogas production**	Grenoble Alpes Métropole	Reduction of mud volumes and needs in gas; Transformation of biogaz into biomethane	Development strategy; Economic model	Partiership; Innovation		Reduction of induced emissions and avoidance of rise in emissions	€€€€	★★★
2	**Geothermal heat**	Eau de Paris	Heat-pump on a reserve of underground water	Partnership for heat sale	Change of paradigm		Increase in avoided emissions	€€€€	★★
3	**Decarbonation**	Suez – Climat Mundi	Decarbonation of drinking water with lime	Communication on the impact of limestone residue on household consumption	Behavioral analysis; Raising awareness towards bottled-water consumption reduction		Increase in avoided emissions	€€€€	★★★★
4	**Low carbon reagent**	Eau de Paris	Replacement of materials and filters	Acquisition policy	Research partnership		Reduction of induced emissions	€	★★★★★
5	**Energy savings**	Eau de Paris	Actions of consumption reduction	Energetic efficiency plan; Financing by the CEE	Certification procedure for ISO 50000-1		Reduction of induced emissions	€	★★★★
6	**Hydraulic optimisation**	Grenoble Alpes Métropole	Modification of the decantation facilities	Elimination of reagents	Research partnership		Reduction of induced emissions	€	★★★★
7	**Pretreatment optimisation**	Grenoble Alpes Métropole	New compactor; Valorization of the grease in digestion	Transportation reduction strategy	Research partnership		Reduction of induced emissions	€€	★★★
8	**Non-collective sanitation**	Aquatiris – Alternative Carbone	Local phytodepuration without all-water pits	Comparison of non-conventional solutions	Engages local populations		Reduction of induced emissions	€	★★
9	**Heat from wastewater**	Veolia - Issy Energies Vertes	External heat exchanger to avoid clogging	Partnership for heat and cold sales	Adoption of new energy services leveraging wastewater heat		Increase in avoided emissions	€€€	★★★★
10	**Inciniration optimisation**	Le Havre Seine Maritime - Veolia	AI-based real time regulation	Flow management to avoid down times	Digital competencies development		Reduction of induced emissions	€	★★★★
11	**Optimized pumps**	SAUR - Usines d'eau potable	Energetic optimisation of the pumps	Planification of renewals	Digital innovation		Reduction of induced emissions	€	★★★★★

Warning! This table is indicative only. When comparing projects, the following must be taken into consideration:

Greenhouse Gas emissions reduction potential :
- a distinction must be made between projects that avoid emissions by third parties and those that enable to reduce the emissions induced by the activity.
- the energy efficiency plan of Eau de Paris is a voluntary engagement tool and its real impact is difficult to quantify

Investment costs :
- the cost indication relates to investments only. Project may have a high or low return rate depending on the sales and savings associated to the operation.

Replicability :
- all the projects accounted for are not on the same scale (big differences in facility and city sizes)
- The phytodepuration solution is a decentralised sanitation solution, which cannot be compared to a large city, centralised sanitation system.

Appendix **Glossary**

Risk and climate change

Adaptation: This is the anticipation of the impacts of climate change and the adjustment process to reduce the vulnerability of natural and human systems to these impacts.

Impact: Impact refers to the effect of a climate hazard on natural and human systems. These effects manifest themselves in a localized way on people's lives, livelihoods, health, ecosystems, economic, social and cultural heritage, services and infrastructure. In this sense, the terms «consequences» or «impacts» are also used.

Mitigation: Actions aimed at limiting the extent of climate change by reducing direct and indirect greenhouse gas emissions. For example, reducing fossil-based energy consumption or reducing N_2O emissions from wastewater are mitigation actions.

Vulnerability: Vulnerability characterizes the propensity or susceptibility of a system to be damaged. It encompasses a variety of concepts, including sensitivity or fragility and the inability to cope and adapt. It therefore depends on multiple factors: socio-economic inequalities, urban development of the territory, implementation of adaptation strategies, etc. It is therefore linked to the political choices and strategies developed locally.

Greenhouse gas and carbon emissions

Carbon footprint assessment: GHG assessment on scopes 1, 2 and 3, based on the emissions at each stage of the system studied and the life cycle of the products purchased or produced.

Emission factors (or greenhouse gas removal factors) **(EF):** this factor is used to transform a physical activity data (activity data) into a quantity of CO2 equivalent.

Greenhouse gas (GHG): a gaseous constituent of the atmosphere, natural or anthropogenic, that absorbs and emits radiation of a specific wavelength in the infrared radiation spectrum emitted by the Earth's surface, atmosphere and clouds. The main greenhouse gases are carbon dioxide (CO_2), methane (CH_4), nitrous oxide (N_2O) and ozone (O_3).

GHG Avoided emissions: these GHG emissions have to be accounted for separately from scopes 1, 2 and 3, as they relate to a reduction in GHG emissions outside of the utility, through the use by a third party of a process by-product. They are calculated based on a reference scenario that needs to be documented. E.g. for water and wastewater services: providing electrical or thermal energy to a third party to use as renewable energy in place of a fossil fuel based energy, providing biosolids to farmers as an agricultural amendment in place of chemical fossil-fuel-based fertilisers.

GHG Emissions Assessment: evaluation of the total quantity of greenhouse gas (GHG) emitted into the atmosphere over a year by the activities of an organization, expressed in equivalent tons of carbon dioxide.

GHG Emissions Scopes: Three categories of emissions are distinguished within the utility boundary: direct GHG emissions (Scope 1), indirect energy-related GHG emissions (Scope 2) and other indirect GHG emissions (Scope 3).

- **Direct greenhouse gas emissions (scope 1):** GHG emissions from sources owned or controlled by the organization. (associated with emission category 1 or scope 1)
- **Indirect energy-related greenhouse gas emissions (scope 2):** GHG emissions from the production of electricity, heat or steam imported and consumed by the organization.
- **Indirect energy-related greenhouse gas emissions (scope 3):** GHG emissions associated to the products and services purchased, sold or disposed of by the utility, including their life cycle.

Water, sanitation and stormwater services

Stormwater Management: The set of measures taken by humans to better control water flows generated by rainfall and runoff in residential areas.

Water and Sanitation Services: The water and sanitation services include the urban water cycle, i.e. all the stages from the collection of raw water, the production of drinking water, the distribution to the user until the discharge of the treated water after it has been purified in its natural environment. In France, public water and sanitation utilities are responsible for the sustainable management of drinking water resources and for the treatment and collection of wastewater and rainwater before it is returned to the receiving environment. The agricultural sector and industries also play a role in preserving the quality and quantity of water resources.

PUBLISHING

Co-published by IWA Publishing,
Unit 104-105, Export Building, Republic,
1 Clove Crescent, East India,
London E14 2BA, UK

Tel. +44 (0) 20 7654 5500
Fax +44 (0) 20 7654 5555
publications@iwap.co.uk
www.iwapublishing.com

ISBN: 9781789063172 (eBook)
DOI: 10.2166/9781789063172

Implemented by:

IWA Publishing's authorised EU representative for General Product
Safety Regulations is Diane D'Arras, 15 rue Duret, 75116 Paris,
France, e-mail: safety@iwap.co.uk.

Printed and bound by CPI Group (UK) Ltd, Croydon, CR0 4YY
12/05/2026
02108881-0019